Praise for *Facing the Climate Emergency*, Second Edition

This book is a guide through the difficult emotions of climate activism, and we are lucky to have it. It helped me understand how to cope with the enormity of the challenges we face, and I bet it can help you too.

—Vanessa Nakate, climate activist, founder, Rise Up Movement, UNICEF Goodwill Ambassador

If you are filled with anxiety and reeling from the onslaught of bad climate news, do yourself and your loved ones a favor and read this book. There is no way around the crisis we face, and no way around the dread we feel except to gather up our courage and fight like hell for the Earth's future. I'm transformed and you will be too.

—Abigail Disney, Emmy-winning documentary producer and director, philanthropist, activist

Facing the Climate Emergency brilliantly addresses a major gap in the movement space—a need to incorporate the psychological needs of the people who want to act, but are struggling to find their path. This book is precisely the antidote to hopelessness and despair that we need. I hope everyone reads it—our capacity to avert a dark future depends on it.

—Jamila Raqib, executive director, Albert Einstein Institution, strategic nonviolent action specialist

This is a life-changing book. Full of tough love, no nonsense moral clarity, and brave determination, Margaret Klein Salamon's words will get you to sit up, reassess your life, and realize the history-making moment you're in.

—Britt Wray, PhD, author, *Generation Dread*, Planetary Health Fellow, Stanford University School of Medicine

This is the most powerful, honest, and psychologically astute book on climate change I've ever read. If we humans have a collective death wish, we certainly express it in our pervasive climate denial. Here is the antidote.

—Richard Heinberg, Senior Fellow, Post Carbon Institute, author, *Power*

Our civilization is asleep at the wheel as we speed headlong towards an irrevocable precipice. Margaret Klein Salamon and her Climate Emergency Fund are on the front lines of transformative action. This book is an exhortation and a wake up call. Everyone should read it.

—Jeremy Strong, actor

There are many intervention points to address the climate emergency, but the most important one is within each of us—not by changing our light bulbs, but by owning our power. *Facing the Climate Emergency* is an invaluable roadmap for everyone who wants to move beyond despair and into effective action. Read it and let's get to work!

—Annie Leonard, co-executive director, Greenpeace US

Most of us, most of the time, live in denial. Margaret is uniquely skillful at pulling us into reality, allowing us to live in full awareness of the challenges we face... Margaret Klein Salamon herself has done the difficult emotional work that our present crisis requires, and because of this she has become a sober, prophetic voice that can cut through the din of apathy.

—Paul Engler, director, Center for the Working Poor, co-founder, Momentum Training, co-author, *This Is An Uprising*

The climate and ecological crisis we face isn't really a technical problem, but a psychological one. A psychologist who has had to overcome her own mental and emotional barriers before being able to face up to the horrific realities of the climate crisis, Margaret Klein Salamon is the perfect guide for the millions of people now going through similar things. If words on a page can help us blossom from worried but passive citizens to empowered and active defenders of our own futures, then those words are Margaret's and they are gathered here.

—Dr. Charlie Gardner, conservationist, activist, associate senior lecturer, Durrell Institute of Conservation and Ecology, University of Kent

Facing the Climate Emergency offers relief for climate anxiety. With the skill of a psychologist and the passion of an activist, Salamon helps readers process fear and grief and find their place in the climate movement.

—Raffi Cavoukian, singer, author, Raffi Foundation for Child Honouring

Margaret Klein Salamon has made a vital contribution to our understanding of the climate crisis, how to access key information, how to take action, how to connect the dots, and how to create the global movement that is necessary to save our world. And she does it from a very refreshing and accessible perspective. A must-read for anybody interested in climate.

—Steven Donziger, environmental and human rights attorney

The only self-help book you will ever need.

—Jessica Wildfire, author

Praise for the First Edition

This book will wake you up, no matter how aware you think you are. Better still: it tells you what to do once you are awake. Read it!

—David Wallace-Wells, deputy editor, *New York Magazine,*
author, *The Uninhabitable Earth*

We are in very hot water, and we can still do much about it — those two facts set the stage for Margaret Klein Salamon's remarkable account of how you can become a climate warrior.

—Bill McKibben, author,
Falter: Has the Human Game Begun to Play Itself Out?

A book we can all use at this time of change and instability. With compassion and understanding, *Facing the Climate Emergency* guides us on the inner journey involved in embracing the truths about the climate crisis, processing our upset over its severity, and awakening the love and courage it will take to solve it.

—Marianne Williamson, best-selling author, *A Politics of Love,*
activist, spiritual leader

In Extinction Rebellion we talk about Regenerative Culture. Healing ourselves, and the world. This work is not optional. If we are to play our role in protecting life, each of us must bring our very best qualities to meet this challenge and reconnect to the Earth and to humanity. This wonderful resource skillfully helps us face our most intense emotions about the climate emergency. Read it, use it, and rebel.

—Gail Bradbrook, co-founder, Extinction Rebellion

Margaret Klein Salamon, climate warrior and psychologist extraordinaire, brings her unique experience, wisdom, and compassion to this groundbreaking book, which will help you transform the jagged scraps of your climate pain into the rich compost of climate action.

—Dr. Peter Kalmus, NASA climate scientist, climate truth teller, author, *Being the Change*

Margaret Klein Salamon is one of just a handful of people whose work has changed the way the world thinks about climate change. Today everyone is talking "climate emergency," but when she began her work most were in denial about the true scale of the challenge. Such is the power of truth well told. Reading this book will change your life and it could lead you to help change the future—for billions of people's lives.

—Paul Gilding, former executive director, Greenpeace International, author, *The Great Disruption*

FACING THE CLIMATE EMERGENCY

Second Edition

Margaret Klein Salamon

with Molly Gage

new society PUBLISHERS

This book is for all life,
and those who fight to protect it.

Cover design by Diane McIntosh.
Cover images: © iStock (cloudy sky bobloblaw, foreground by-studio)
Step title images: Step 1 © AdobeStock_35555859; © iStock (Step 2 eugenesergeev, Step 3 whiteson, Step 4 Justin Hobson, Step 5 Xurzon)

Printed in Canada. First printing April 2023.

Inquiries regarding requests to reprint all or part of *Facing the Climate Emergency* should be addressed to New Society Publishers at the address below. To order directly from the publishers, please call 250-247-9737, or order online at www.newsociety.com.

Any other inquiries can be directed by mail to

New Society Publishers
P.O. Box 189, Gabriola Island, BC V0R 1X0, Canada
(250) 247-9737

LIBRARY AND ARCHIVES CANADA CATALOGUING IN PUBLICATION

Title: Facing the climate emergency / Margaret Klein Salamon with Molly Gage.
Names: Salamon, Margaret Klein, 1986- author. | Gage, Molly, 1978- author.
Description: Second edition. | "How to transform yourself with climate truth." | Includes bibliographical references and index.
Identifiers: Canadiana (print) 20230165451 | Canadiana (ebook) 2023016546X | ISBN 9780865719910 (softcover) | ISBN 9781550927832 (PDF) | ISBN 9781771423793 (EPUB)
Subjects: LCSH: Environmental degradation—Psychological aspects. | LCSH: Global environmental change—Psychological aspects. | LCSH: Climate change mitigation—Citizen participation. | LCSH: Green movement—Citizen participation.
Classification: LCC GE140 .S25 2023 | DDC 363.7—dc23

Funded by the Government of Canada | Financé par le gouvernement du Canada | Canada

New Society Publishers' mission is to publish books that contribute in fundamental ways to building an ecologically sustainable and just society, and to do so with the least possible impact on the environment, in a manner that models this vision.

FSC www.fsc.org MIX Paper from responsible sources FSC® C016245

Certified B Corporation

new society PUBLISHERS

CONTENTS

Foreword xi

Preface: A New Age of Heroes xiii

Introduction 1

Step One:

Face Climate Truth 15

Step Two:

Welcome Fear, Grief, and Other Painful Feelings 41

Step Three:

Reimagine Your Life Story.............................. 63

Step Four:

Enter Emergency Mode 75

Step Five:

Join the Movement and Disrupt Normalcy 95

Conclusion: All-In for All Life 115

Endnotes 120

Index 131

About the Authors...................................... 137

About New Society Publishers........................... 138

FOREWORD

NEVER HAVE I ROOTED for a book *not* to have a second edition more than Margaret Klein Salamon's *Facing the Climate Emergency: How to Transform Yourself With Climate Truth.* Yes, I understand a second edition usually means a book is a success. And yes, I understand a second run means a multitude of readers connected with Margaret's empathetic yet passionate psychological roadmap to climate action. And yes, the second edition is even more galvanizing than the first, highlighting the bravery and effectiveness of disruptive nonviolent climate activists.

But I was hoping against hope that Margaret's brilliant book would simply... no longer be necessary. And that humanity would have already begun its global transformation toward a fossil fuel-free future. How can we not have?

The science is so incredibly jarring.
The stakes are so high.
The pathway to action is so clear.

And yet nations and industries of the world are practicing predatory delay every day, action has been incremental, and the media expresses the threat in muted terms and quickly moves on. Or even worse, it doesn't express it at all. I'm not a psychologist like Margaret, but I believe the scientific term for all this is *bananas.* The world's gone mad and needs therapy, stat. This book offers help.

As I write this, we are already living on a planet 1.1° Celsius warmer than preindustrial levels. England hit its first 40° Celsius (104° Fahrenheit) day in recorded history this past summer, trees now grow in the Arctic tundra, and the massive Thwaites ice shelf will likely break apart in less than five years—all thresholds not expected to be crossed until 50 to100 years from now. And yet we still hedge, dither, and delay.

It's a frightening, maddening, and extremely dangerous situation. So, thank God for this second edition of *Facing the Climate Emergency*. Because, man oh man, do we need to face the climate emergency. Now.

I am proud to support brave, nonviolent, disruptive activists who are helping to wake up the public through Climate Emergency Fund, where I am a board member, and Margaret is the executive director. This book explains why and lays out clearly how to take the necessary journey from fear and despair to action and vision. Also, all royalties from this book go to Climate Emergency Fund, so thank you for joining me in supporting the climate movement's vanguard!

Ultimately Margaret Klein Salamon understands, as few do, that the biggest challenge we face with the collapse of the livable atmosphere is that it is almost emotionally unimaginable in its scope. And that until we approach the immensity of the challenge that we face, we can't begin to make it smaller.

So, I hope this book becomes as ubiquitous as Heimlich maneuver posters in restaurants. I hope the breadcrumbs it lays for us to navigate this new warming reality become as common sense as not smoking on an airplane or wearing a seatbelt. But most of all, I hope when a third, fourth, or fifth edition of this book is discussed, someone raises their hand and says, "Why would we do that? Who doesn't understand what we're facing?"

Adam McKay, Writer and Director, Don't Look Up

PREFACE: A NEW AGE OF HEROES

When I wrote the first edition of *Facing the Climate Emergency* in 2019, the idea that the climate emergency was profoundly affecting our psychology was still a bit marginal. But conditions are changing rapidly.

Today, the pain is more widely acknowledged. The world's largest-ever study of young people's emotional reactions to the climate emergency surveyed 10,000 in 10 countries. Published in 2021 in *The Lancet Planetary Health*, this study found that 56 percent of young people believe that humanity is doomed, and 46 percent report that climate distress affects their daily ability to function. They blamed their governments: 58 percent reported feeling betrayed by governmental inaction. According to another survey from 2020, more than a third of Millennials and members of Gen Z don't want to have children because of climate change.[1]

But we are also in a new age of heroes. More and more people, young and old, are going all-in for all life, and they are doing extraordinary things. You will hear from several of these activists below. They are smart, strategic, brave, and desperate. They are channeling their feelings into a powerful movement that demands rapid global transformation of our energy, agricultural, economic, and social systems. As movement historian Sarah Schulman puts it, "People who are desperate are much more effective than people who have time to waste."

These heroic activists can spark the collective awakening we need. Indeed, it's already happening. These activists are our best hope, and I will be highlighting their work throughout this text. My goal in this book is to inspire you to take your rightful place beside them. Beside us, I should say, as I am proud to play a supporting role in the movement.

Since the first edition of this book came out in 2020, billions of people were infected by COVID-19, Russia invaded Ukraine, and so many floods, droughts, storms, and heatwaves broke records that I will not endeavor to catalog them.

These factors together have caused hunger to rise dramatically. A recent UN report stated that 2.3 billion people in the world were moderately or severely food insecure in 2021, and 11.7 percent of the global population faced food insecurity at severe levels.[2]

The climate emergency, already wreaking havoc on the food supply, is accelerating. We are in the endgame now, and we bear a tremendous responsibility in this moment. I hope we can work together to meet it. Onward!

INTRODUCTION

> *Not everything that is faced can be changed. But nothing can be changed until it is faced.*
>
> —James Baldwin

We are in a moment of collective suffering. As the accelerating climate emergency devastates homes, crops, and lives, primarily in the Global South, it is contributing to despair and anxiety everywhere. In the United States, deaths of despair are increasing. Suicide rates rose 30 percent between 2000 and 2020.[1,2] Opioids kill more Americans than car crashes.[3] One in six Americans takes psychiatric medication, primarily for depression and anxiety.[4] Virtually all of us resort to numbing or distraction: We watch 33 hours of TV a week, scroll endlessly on social media, play video games, and watch pornography.[5] We drink too much, eat too much, work too much, compete too much, and buy too much. Simply put, Americans—and people all over the world—are in pain.

There is, of course, an enormous self-help industry dedicated to helping us feel better. Books, podcasts, and seminars say we're unhappy because we are disconnected from other people and over-connected to technology; they say that we are harshly self-critical and care too much about what other people think; that we live in a suspended adolescence; that we don't manage our money properly; don't take enough emotional risks, aren't living our best lives, and don't practice enough self-care. Others rightfully point to the glaring realities of inequality, poverty, and precarity as the cause of our pain. These narratives are each important and valid to some degree, and they may offer some help.

But there's something else going on; something is eating at us. We are in pain because our world is dying, and through our passivity, we are responsible for killing it.

Inside all of us, a battle rages. It's the battle between knowing and not knowing, between fully facing the truth—emotionally and intellectually—and shrinking from it. We sense we're in a climate emergency and mass extinction event, but we have a deep-seated psychological instinct to defend against that knowledge. The pain is shouting at us: "Everything is dying!" Part of us knows that humanity and the natural world are in peril. Indeed, we feel the horrors of civilizational collapse and the sixth mass extinction of species in our bodies. But another part attempts to shield us from this pain—we avoid and deny, distract and numb ourselves from what we know. These defenses work, but only temporarily: When we fail to process our emotions and mourn our losses, the pain takes on tremendous power. It follows us around like a shadow, and we become increasingly desperate to avoid the by now obvious truth.

This pain has several dimensions. It is the fear we feel for ourselves, our loved ones, and for all humanity; it is the empathy and grief we feel for the people and species already immiserated or killed; it is the crushing guilt we feel for continuing to let this happen. Our pain is the consequence of our participation in a destructive system. We have allowed ourselves to become killers—a plague on the rest of life. We share, to varying degrees, guilt and responsibility.[6,7]

Our pain may feel terrible, but it is rational, appropriate, and deserved. It is an internal reflection of external reality: The biosphere—all life—is suffering or threatened. *Of course* we feel sad and anxious. We are caught in an economic and political system that fosters our collective participation in our planet's daily degradation. Why would we expect to feel good, or good about ourselves while we participate in killing all life on Earth, including ourselves and everyone we love?

On the one hand, we are victims. No one asked to be born into this broken system that treats all life as disposable and allows for unprecedented levels of inequality. We have been failed by the people and institutions tasked with protecting us—first and foremost, our governments and elected representatives. This governmental failure could not be more complete. The total abdication of duty to protect humanity and all life has made the social contract between government and citizens a sick joke.

But the government is not alone: Media outlets, universities, churches, museums, labor unions, environmental organizations, professional associations, and countless others have also failed to acknowledge and protect us from the climate emergency.

And, of course, corporations, such as fossil fuel companies like ExxonMobil, bear the lion's share of responsibility for the global cataclysm that is well underway. For decades, the fossil fuel industry has run a multibillion-dollar campaign of lies and climate denial. It has successfully sowed doubt in our society and blocked anything approaching an appropriate response from our elected leaders. The level of cravenness required to lie to the public about catastrophic warming to continue our addiction to fossil fuels is appalling.

Many other corporations are implicated. Monsanto and other agricultural corporations, big banks, airlines, carmakers, and others have pursued a similarly environmentally devastating business model—killing and endangering life in exchange for short-term profits.

On the other hand, we—you and me—are not merely victims. Through our passive participation in this system, we are also perpetrators. We have failed ourselves and each other. We've allowed our home to be robbed, and now we are watching it burn. Although humanity has become almost godlike in our power to create and destroy, we have remained childlike in our use of that power. It's time to find our maturity and our heroism.

In 1956, psychoanalyst and activist Erich Fromm wrote *The Art of Loving,* which examined the psychological impacts of a consumer-capitalist society on individuals. Fromm argued that people are alienated from their work, from themselves, and from each other. He noted that people had been sold the view that life was one big competition or marketplace and that people were commodities who should try to maximize not just their money but also their popularity and attractiveness. He observed that people in these societies treat themselves like commodities in a competitive market, adopting false selves to fit in and be liked while abandoning their authenticity and sense of true purpose.

This ideology still prevails today and fosters the following beliefs:

- You are an isolated individual, defined by what you achieve, how much you earn, and what you buy.
- You should focus on competition with others and personal indulgence.
- The only way to receive love and acceptance is to own more things.
- There is no community, and there is no web of life.
- Other people are threatening, especially people who are a different race or culture.
- You have no moral responsibilities. You are a deprived victim who deserves much more than you get.
- You are living at the pinnacle of human achievement, defined by constant economic growth, and it's naïve to think there could be anything different.
- You may feel unpleasant feelings, but they will disappear if you buy something.

In his essay "Love of Death and Love of Life," Fromm postulated that the only reason people would not rise against the possibility of worldwide nuclear destruction was if they were already anticipating that destruction: On some level, he reasoned, the destruction must have felt appropriate and even appealing—better, at least, than this bullshit, dead-end, alienated, and humiliating life. Otherwise, why did our society allow the risk of mass nuclear obliteration to threaten us for decades? Fromm believed if people inherently felt their lives were precious and worth living, if they felt engaged in life and saw that engagement reflected in others, if people were not housing a deadness within, they would demand an end to the creation of weapons of mass destruction. They would refuse to accept the possibility of the end of all life.[8]

When it comes to the climate crisis, we must ask ourselves the same question: When faced with our current ecological disaster and the worst that is still to come, why are we passively accepting mass suicide and mass murder? Have we given in to death's pull?

When we see the media address dire scientific reports in a few stern sentences before cutting away to celebrity gossip, when we see passivity and resignation to our fate from friends and community members, when we hear the refrain of "we're fucked," we have to conclude that the coming ecological crisis must feel like an expected and maybe even a fitting end to our degraded society. How can we otherwise make sense of the fact that more people aren't rioting in the streets at the imminent destruction of their lives, their children's lives, and the entire web of life?

Our society treats life—human, plant, and animal life—as if it were a cheap commodity rather than the most precious, sacred thing there is. By doing so, we've not only created the ecological crisis, we've desensitized ourselves to it. Maybe this is why worldwide annihilation seems to be an appropriate end: It reflects the emotional and spiritual destruction we've internalized.

Do we want to live? If we do, we need to wake up and grow up—right now. We may be about to lose everything, but we aren't dead yet. It doesn't have to be this way. We can face climate truth and choose not to commit passive suicide.

Indeed, we are in a new age of heroes. More and more people are going all-in for all life, and they are doing extraordinary things. You will hear from several of these activists below. They are smart, strategic, brave, and desperate. They are channeling their feelings into a powerful movement that disrupts normalcy and demands rapid global transformation of our energy, agricultural, economic, and social systems. These activists can spark the collective awakening we need. They are our best hope. My goal in this book is to inspire you to take your rightful place beside them.

We can choose to turn away from illusion and distraction. We can each decide to face climate truth and decide that now is the time to do everything in our power to wrest life back from the jaws of extinction. We can each help to initiate a collective awakening to the climate emergency and a World War II-scale response that protects humanity and the natural world and builds a beloved community.

One of my deepest beliefs is in the immensity of human potential. As a therapist, I have seen people overcome trauma, addiction, and personality issues—and thrive. Looking at the history of social movements, I see how society evolves by leaps and bounds. I see how, during all-hands-on-deck emergency mobilizations, the impossible becomes possible. During WWII, the first computer came into regular use, RADAR and blood transfusions were innovated, as well as breakthroughs in manufacturing, class, race, and gender equality. For a few years, taxes on the highest earners were over 90 percent, and factories had daycares and takeaway meals for women who joined the workforce in droves.[9]

Today we need a national emergency mission of a comparable scale to restore our climate and natural world. But we can do it! In this, my beliefs align with Oscar Wilde: "The only thing that one really knows about human nature is that it changes."[10]

Most people approach climate in terms of limiting damage, but we can aim higher than that. We can set the goal of restoring biodiversity and the preindustrial climate.[11] I refuse to give up on a vision of an Earth teeming with life. Some will say it's impossible, that too much damage is already done. But we have barely scratched the surface of human potential. We don't know what our limits are in protecting our climate because we have never tried. Once we actually get started, with all hands on deck, directing our collective brilliance toward responding to this emergency, we will do amazing things. Humans will finally act as responsible stewards of nature, and of each other.

To do so, we must shake off our resignation and our selfish consumerist programming and denounce the lies that aid and abet denial. Each of us must do our part to reestablish our connection to humanity and all life, and to recognize our bottomless responsibility to protect it. We must acknowledge that responsibility, and use our talents, energy, privileges, and resources in the struggle. We must join with each other. We must allow ourselves to face the truth and to accept the reality that we must transform—now—individually and together, to respond effectively to the climate crisis.

Socialism has experienced a resurgence in recent years—partly because many people see capitalism as responsible for the climate emergency. It's true: Capitalism, with its dependence on endless growth, its tendency to concentrate wealth and increase inequality when unchecked, its treatment of workers as disposable and the living world as expendable, and its relentless use of advertising to make good citizens synonymous with good consumers, is fundamental to the problem.[12]

However, we can't blame capitalism alone. Our industrial and extractive approach to meeting human needs is bigger than capitalism, which is simply the main scheme by which these destructive activities are coordinated and their ill-gotten gains hoarded. The communist totalitarian Soviet Union was easily as damaging to ecosystems as the market-based United States[13] and was the second-largest emitter of greenhouse gasses during the 1960s, '70s, and '80s.[14] China, with its primarily state-driven economy, has now become the world's largest emitter of greenhouse gasses. The social democratic state of Norway owns 67 percent of Equinor, formerly Statoil—an oil and energy company.[15]

To have any hope of surviving the climate and ecological catastrophe, we must transform our destructive economy into a regenerative one, and do it at emergency speed. We don't just need zero emissions in every sector; we need massive carbon drawdown projects that restore ecosystems and the soil. We need permaculture and food localization; we need an end to mass consumerism and endless growth; we need to give back half the Earth to nature to restore biodiversity,[16] and we need to create a society based on protecting and healing humanity and the natural world. This means transforming not only our energy, agricultural, transportation, and industrial systems but also transforming ourselves and how we relate to each other. We need to rethink our basic concept of who we are and what matters.

And we need to do it all right now.

This is a self-help book, but its goal is not to make you feel less pain. Its purpose is to make you feel your pain more directly and constructively, to turn it into action that protects humanity and all life. I argue that your pain is a signal—it's telling you something critically important. The pain is demanding to be acknowledged, and I want to show you how to listen and learn how to attend to it. I want you to face the pain of the climate and ecological emergency and to feel it in a focused, conscious way so that the pain launches you into a process of transformation—first in yourself and then in society as a whole. This large-scale change must be our goal, as Pope Francis wrote in his 2015 encyclical Laudato Sí'. To stop the climate emergency, he says, we must "become painfully aware, to dare to turn what is happening to the world into our own personal suffering and thus to discover what each of us can do about it."[17]

I've been through this process myself: I've felt the pain, faced it, and I have been transformed by it. Ten years ago, in 2012, I was a young professional in New York City—a clinical psychologist working on a doctoral degree, preparing to enter private practice and start paying off my six-figure student debt. I avoided thinking or reading about the climate because it made me feel terrified and helpless. I would read the first sentences of articles about global warming, say to myself, *Nope! I can't handle it; it's too scary.* Then I'd close the article and distract myself with something else.

Then Hurricane Sandy hit New York City, and everything came to a standstill. Destruction was everywhere. I vividly remember seeing a car smashed by a huge branch. On the shattered windshield, a cardboard sign read, *Is global warming the culprit?* Seeing the message caused something in me to shift. I knew the answer to that question, though my knowledge was diffuse. That sign helped me focus: If global warming had smashed that car and the whole city, what else could it do? How bad was this situation, and what did our collective future hold? With these questions in mind, I started to educate myself. This was the beginning of my waking up. I began to finish the articles that had previously overwhelmed me. I started to seek out books on the climate and ecological emergencies.

What I learned shook me to my core—and caused me to reassess my life. I realized that it was my responsibility to do everything I could to halt and reverse the coming catastrophe.

So I left the field of clinical psychology—which I love—and dove headfirst into activism. My husband and I focused on necessities, moving into a small apartment and making the minimum student loan payments, so that I could build an organization as a volunteer. He supported us, and my parents helped out as well, privileges relatively few have.

First, I published a blog—*The Climate Psychologist*—which was oriented toward recruiting like-minded people to build a social movement. I met Ezra Silk through that blog, and we founded and built an organization called The Climate Mobilization, or TCM. The Climate Mobilization was a combination of a think tank and an advocacy organization. We published cutting-edge thought leadership advocating for WWII-scale climate mobilization.

With the help of an amazing team of volunteer organizers, we mainstreamed the "Climate Emergency" frame by initiating a city-based Climate Emergency declaration. This strategy was operationalized on a city-by-city basis: Hoboken, New Jersey, Montgomery County, Maryland, and Berkeley, California were among the first cities to declare a Climate Emergency after vigorous local organizing campaigns. We expanded its reach by sharing the campaign with Extinction Rebellion and watched as it went viral. In 2019, use of the term "Climate Emergency" went up 10,000 percent and *Oxford* declared it Word of the Year.[18]

As the broader movement took up the ideas of Climate Emergency and WWII-scale Climate Mobilization, I began to focus on two areas: the emotional aspects of the climate emergency and strategic philanthropy. I published the book's first edition in 2020 and then launched Climate Emotions Conversations, an online platform through which hundreds of people from all over the world have shared their climate feelings in small-group guided conversations.[19]

Usage of "Climate Emergency"

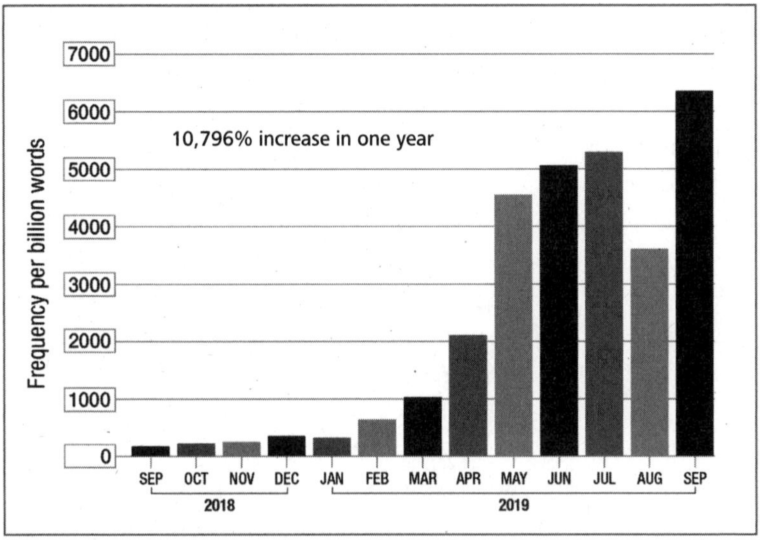

Through this, and my other work in the movement, I've had countless conversations with people—from elected leaders to climate scientists to stay-at-home moms to hedge fund managers to janitors—about how to process fear and respond to climate truth.

I was an advisor to Aileen Getty, Rory Kennedy, and Trevor Neilson as they set up the Climate Emergency Fund in 2019, and in 2021, I joined as Executive Director. It is an honor to fundraise and make grants to the climate emergency movement.

My climate mission is a light inside me, it's my animating force. I want that feeling for you, too. As Albert Camus put it, "In the midst of winter, I found there was, within me, an invincible summer."[20] The winter of climate catastrophe is everywhere, and yet through living in climate truth and dedicating yourself to protecting humanity and the living world, you will stay warm and well.

Since I faced up to the truth and started to change my life and support the movement, my life has improved in countless ways. I used to be harshly self-critical. But knowing that I am doing everything I can for the mission has helped me find a sense of inner peace. I am obsessed with it. It is such a relief to care about something more than myself.

This may sound like an outlandish mentality and lifestyle, but I see people undergo this transformation all the time. It's what I call "the movement mentality." All over the world, ordinary people are performing extraordinary feats of service—rearranging their lives, leaving their jobs, donating until it hurts, and risking arrest, to go all-in for this mission. When asked about their motivations, activists frequently say something along the lines of, "this is all that matters," or "I don't want to be doing this, I have to."

It's not easy, and I'm not perfect. I'm always busy, frequently stressed, sometimes grumpy, and occasionally mean. I have more student debt than when I earned my doctorate. I sometimes neglect my health. But I wouldn't want to live any other way. I would never, ever go back to my state of passive ignorance.

As the number of people operating with the movement mentality increases, so will the success and urgency of the movement. We are seeing the rise of a new, uniquely powerful, focused, and determined, disruptive global activist movement.

At Climate Emergency Fund, I have had the honor of supporting some of the leading disruptive climate activism in the world: Climate scientists chained themselves to the White House and the doors of banks in order to communicate the urgency and wake up the public, Just Stop Oil blocked eleven fossil fuel infrastructure sites at a time, impacting the oil supply to whole regions of the UK[21] and famously threw soup onto paintings. [22] Meanwhile their sister organization, Dernière Renovation, disrupted the Tour de France.[23] These are all spectacular nonviolent, disruptive actions calculated to wake up the public.

It's working. This is what the beginning of a collective awakening looks like—more and more people deciding that they will not sit quietly while the world around us dies. The awakened must also make it our job to wake up the rest. The members of this movement are not content to numb our sadness with money and things. We're not willing to ignore the Earth as it burns. We're going to fight like hell for everything we love.

This book will show you how to join our ranks as members of the climate emergency movement. In it, I will ask you to tap into your fear about our current climate crisis and the future we are careening toward. I will help you mourn what has already been lost and what we continue to lose every day. I will help you transform your despair into a collective effort to build power for the movement.

It's not going to be easy. It's going to be the opposite of easy. But acknowledging the truth of our climate and ecological emergency, grieving our lost futures, and taking the heroic path will make you confident and strong. It will give you a mission and purpose beyond anything you have experienced. It will allow you, at long last, to heal your pain and feel genuinely good about yourself.

It will connect you to your fellow humans, and it will connect you to all life. It will give you real hope, based on your real potential to affect real change.

But most importantly, it will help give humanity a better chance of canceling the apocalypse and protecting itself and the living world.

It will be a difficult journey, but I can promise you that something extraordinary will happen when you commit to it. You will feel hope; you will know that you are part of the solution; you will see that you are doing your part to save the world. In other words—it feels fantastic to contribute to the movement!

When you face climate truth and let it transform you, you will become heroic, leveraging your talents, energy, and resources in service of protecting humanity and all life. No one is coming to save us, but together, we might be able to save ourselves.

Questions for Reflection and Discussion

- How do you feel about the climate emergency?
- Do you feel anxiety, depression, or have other painful psychological experiences that you struggle to define?
- How might the climate crisis and species extinction drive some of those feelings?
- Have you experienced any sense of the inner deadness that Fromm describes?
- How and to what extent have you replaced your love of life with a love of objects?
- Can you imagine yourself as a hero? As a protector?

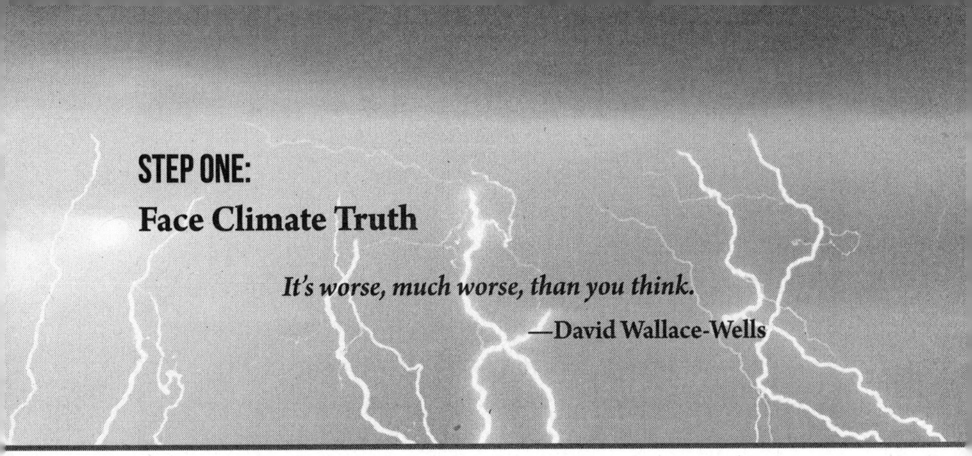

STEP ONE:
Face Climate Truth

It's worse, much worse, than you think.

—David Wallace-Wells

Have you noticed how good, freeing, and fortifying it feels to tell the truth, and how heavy it is to carry around secrets or lies? Therapists know people create transformative change by reckoning with reality. That's why the first step toward transformation is to face climate truth with your whole self—with both your intellect and your emotions, both your mind and heart.

Some people rationalize their lack of engagement with the climate emergency, telling themselves that they don't have the relevant expertise to understand the situation, that only scientists can understand it. But each one of us is fully capable of understanding and responding to climate truth.

When someone we love is diagnosed with cancer, we don't say to each other, *Well, okay, this is an issue for science to solve,* and turn away. Instead, we feel intense fear, pain, anticipatory loss, and grief, and we look for ways to help. We know that, for the cancer sufferer, integrating the diagnosis will be a challenging emotional process. We also know that when the cancer sufferer denies or doesn't emotionally process their illness, it's a serious problem that can impair treatment.[1]

Just as we can grasp the far-reaching impact of a cancer diagnosis without being doctors, we can grasp the impact of our climate emergency without being climate scientists. To respond fully and humanely to the climate emergency, we need only understand the basic concepts of the crisis and its implications, allow ourselves to face and feel the feelings we're avoiding, and then act.

In two sentences, the truth of the climate emergency is:

> Scientific consensus says the accelerating climate emergency and ecological crisis threatens human life through famine, disease, accelerating natural disasters, conflicts, and social collapse. Only an emergency mobilization of resources to rapidly transform our entire economy and society can possibly protect the future of humanity and the living world.

The reality of the current and coming catastrophe can feel overwhelming, and many of us turn away, telling ourselves that we can't handle it or are just not ready. This is an understandable response—I've had it myself—but it will not protect humanity or the natural world.

So, don't ask yourself whether you are *ready* to face the reality of our climate emergency or whether you can handle it. Instead, ask yourself, *What is my priority*? Would you rather protect yourself from painful knowledge or would you rather protect yourself, your family, and the entire human family from the actual climate emergency? These are your two options. The choice should be easy.

Furthermore, avoiding the truth requires constant effort and vigilance. You may not even realize how much energy you spend *not* acknowledging environmental collapse. It is hard work to constantly protect yourself from painful knowledge, to not let yourself feel your fears. When you avoid the truth, you put the energy that could be used toward *preventing* the climate emergency into safeguarding the fiction you've created for yourself.

Believe me, it is an incredible relief to let go of your defenses, vigilance, and denial, and allow yourself to live in truth. Integrating climate truth into your life will make you more capable of feeling joy and facilitating change.

Facing climate truth means recognizing, as journalist David Wallace-Wells writes in the opening lines of *The Uninhabitable*

Earth, that "it's worse, much worse, than you think."[2] It means realizing that many parties are creating an unrealistically optimistic picture of the climate emergency. Contributors include the fossil fuel industry's misinformation campaign, the failure of both political parties to reckon honestly with the emergency, the Intergovernmental Panel on Climate Change's (IPCC) systematic bias toward understatement, the misguided euphemisms of the gradualist climate movement, and our psychological defenses.

Thanks to American scientist Eunice Foote and Irish physicist John Tyndall, we've known since the 1850s that carbon dioxide (CO_2) is a heat-trapping gas.[3] In 1898, Swedish scientist Svante Arrhenius identified the global greenhouse effect, theorizing that burning fossil fuels could cause the atmosphere to warm.[4] Additionally, English engineer Guy Callendar demonstrated in 1938 that the Earth was already warming due to carbon dioxide emissions. However, it took a couple of decades for his work to gain scientific acceptance.

The facts about CO_2 and the greenhouse effect were not limited to academics. ExxonMobil's scientists have been aware of the greenhouse effect since the 1970s.[5] They even had a team of top scientists conducting pioneering research into the problem. But rather than using that information to inform the public of the existential risk and transitioning their business so that it would not destroy civilization and the living world, ExxonMobil launched a massive disinformation campaign to create doubt about the scientific certainty of the greenhouse effect.

In *Merchants of Doubt* (2011), Naomi Oreskes documents how the fossil fuel industry followed the "tobacco strategy" to cast doubt on settled science.[6] Amazingly, some of the same scientists, such as Fred Singer, who fought against the scientific evidence that smoking causes cancer, spent the later years of their career manufacturing doubt around the settled science that fossil fuels are warming the planet.

Fossil fuel companies paid scientists like Singer and Willie Soon and supported denialist think tanks, such as the Nongovernmental International Panel on Climate Change (NIPCC). The Heartland Institute spread the NIPCC's false claims. To extend the campaign's reach, they targeted America's teachers with misinformation. The Heartland Institute mailed a book and DVD, *Why Scientists Disagree About Global Warming,* to 350,000 American public school science teachers in 2017.[7] Not surprisingly, the fossil fuel industry has also utilized social media in their campaign of lies; in 2021, sixteen of the world's biggest polluters spent at least $5 million to spread false and misleading content on Facebook through ads that generated around 150 million impressions.[8] They also bolster their legitimacy by sponsoring museum exhibits, donating hundreds of millions of dollars to universities for research, and advertising in the *New York Times.*[9]

This campaign of lies and political manipulation has cost fossil fuel companies hundreds of millions of dollars, and it has been remarkably successful.[10] If you have ever had any questions or doubts about whether the greenhouse effect is a scientific consensus, it's because the oil industry's campaign worked. If you have ever allowed yourself not to worry about the climate emergency because "scientists are still figuring it out," it's because of the skepticism the industry successfully cast on basic science through misinformation. If you have ever held yourself back from discussing the climate emergency because "it's too controversial," it's because the oil industry has strategically invested in your silence.

At the same time that the fossil fuel industries were building a disinformation campaign to feed our defenses, fossil fuel magnates David and Charles Koch and other fossil fuel interests were successfully preventing our government from taking the climate emergency seriously. Fossil fuel companies spent nearly $2 billion lobbying against climate legislation between 2000 and 2016.[11, 12] The GOP is the only political party in the world that continues to deny the science of global warming.[13] And for decades, the Democratic Party has drastically understated the threat and the scale of response required to protect humanity and the living world. Trump was an exaggerated embodiment of the firmly entrenched, bipartisan commitments to ignore and underplay the crisis.

Presidential administration after administration has failed to reckon with the scale of the crisis. In the early 1990s, George H.W. Bush's words at the 1992 Earth Summit—"The American way of life is not up for negotiation"—laid a strong foundation for administrative support of wanton consumerism.[14] Obama boasted about increasing the United States' oil production "every year I was president," turning the United States into the world's largest exporter. "That was me, people," he reminded the crowd in Houston, Texas, "say thank you."[15] This was in 2018, as the climate was spinning out of control.

In 2022, the Democrats finally passed major climate legislation, known as the Inflation Reduction Act, marking a crucial new phase in the fight for real climate action. Independent analysis suggests this bill will, alone, get the United States 40 percent below 2005 emissions by 2030. It is likely to kickstart a clean energy, agriculture, and transport revolution, and the importance of that momentum-building can't be overstated.

This bill would not have been possible without climate emergency activism. Extinction Rebellion, Sunrise Movement, School Strikers, Indigenous water protectors and pipeline fighters, and their many allies and coalition partners who led the way, deserve credit. In the runup to the Inflation Reduction Act (IRA), Climate Emergency Fund supported activists who held a hunger strike in front of the White House and protested Senator Manchin at his Maserati, at his houseboat, and at the coal plant in West Virginia that enriches him.

While we recognize and celebrate our successes, we must remain firmly rooted in reality. The bill is wildly insufficient.

Because our political system has been captured by big money, this bill didn't do anything to directly stop or penalize the fossil fuel industry, which is still, suicidally, expanding. It does little for biodiversity or land protection. The bottom line? The IRA won't be enough on its own to stop the Earth from hitting catastrophic climate tipping points.

We must escalate our resistance. If we don't, we will soon be living in the Inflation Reduction Act future, where cleaner, cheaper energy, electric cars, and trucks offer a temporarily tolerable lifestyle for the privileged, while electric tanks at the border stop desperate migrants from coming in; where solar-powered air-conditioned indoor farms grow fresh greens and berries for the wealthy, while others choke in dust storms, suffer in heat waves, and starve; and the fossil fuel industry hangs on to wealth and power with increasingly desperate and violent measures. This eco-apartheid has already begun, and we are racing toward total collapse, in which everyone, even the privileged, loses everything.

Futurist and technologist Douglass Rushkoff argues that this is exactly what the ultra-rich are planning for in his book *Survival of the Richest: Escape Fantasies of Tech Billionaires.* He was paid by billionaires to consult on how to survive as society breaks down. They interrogated him about how to maintain authority over their security forces during a collapse:

> They knew armed guards would be required to protect their compounds from raiders as well as angry mobs. One had already secured a dozen Navy SEALs to make their way to his compound if he gave them the right cue. But how would he pay the guards once even his crypto was worthless? What would stop the guards from eventually choosing their own leader? The billionaires considered using special combination locks on the food supply that only they knew.

> Or making guards wear disciplinary collars of some kind in return for their survival. Or maybe building robots to serve as guards and workers—if that technology could be developed "in time."[16]

This is how much of the wealthy elite are approaching the climate emergency—building bunkers and imaging they will be able to somehow maintain control, using violence, while everything collapses.

The denial campaign waged by the fossil fuel industry is so powerful because of an unlikely ally—our own minds. We tend to defend against painful information, and the denial campaign aligns with our desires and defenses. The denial campaign is telling a story we all wish were true. Most of us would prefer not to face the climate emergency and the "inconvenient truth" that our primary power source for electricity, transportation, and manufacturing is deadly. We would rather not think about it, not have an emotional reaction to it, not talk about it, and not recognize solving it as our responsibility. So we use every psychological defense in our arsenal, including:

- Denial: *It's not real.*
- Intellectualization: *It's real, but it doesn't affect me emotionally.*
- Willful ignorance: *I don't want to know what's happening; it's too scary.*
- Wishful thinking: *It can't be that bad.*
- Regression: *We need the experts to handle this.*
- Rationalization: *I can't do anything meaningful to address it.*
- Compartmentalization: *It's irrelevant to my feeling like a good, moral person.*

- Projection: *It's happening, but it's other people's fault and responsibility.*
- Dissociation: *I will zone out or numb myself with substances, video games, or other distractions.*

We also share a common tendency to focus on individual consumption choices. We might say, This is happening, and it's my responsibility to purify my consumption and reduce my carbon footprint. While this is, in some ways, laudable, it can also be a defensive distraction from creating transformative societal change.

Minimization of the emergency is widespread in international climate negotiations, and the mainstream climate and environmental movement habitually understates the risks of the climate emergency. Even groups we think of as on the right side of this issue have failed to tell us the whole truth.

The Intergovernmental Panel on Climate Change (IPCC), the United Nations' body driving much of the climate change conversation, is an inter*governmental* panel, not a purely scientific one. Governments appoint the lead authors of the IPCC's scientific reports, and all members—including petro-states like Saudi Arabia, Russia, and the United States—directly influence the IPCC's *Summary for Policymakers* reports, which drive media coverage on climate change.[17] Climate science and policy analyst David Spratt and former fossil fuel industry executive Ian Dunlop demonstrate how "scholarly reticence" and governmental influence in the IPCC have created a systemic bias toward euphemism and an understatement of existential risk.[18] Even so, the IPCC has been taking an increasingly desperate tone in recent reports.

In 2017, when Wallace-Wells published his wildly popular article on the possible worst-case scenarios of the climate crisis, he was reprimanded from within the climate movement and called a "doomist." World-famous climate scientist Michael Mann wrote in response, "fear does not motivate, and appealing to it is often counterproductive as it tends to distance people from the problem, leading them to disengage, doubt, and even dismiss it."[19]

These comments reflect what has become orthodoxy in the mainstream, or what I call the "gradualist" climate movement: *We must not scare the public; they cannot handle it.* Examples abound. Carbon Tracker published a white paper called "Fear Doesn't Work and Other Lessons from Climate Communication" (2017). In it, and elsewhere, there is often the common refrain that we need *hope,* not fear. See, for example, the more recent development of "climate optimism."

This misguided and counterproductive mandate has its origins in the culture of science, which tends to treat emotion as a threat to rationality. The "fear of fear" is reinforced by philanthropy, which is funded by corporations and the very rich, who generally prefer cheerful optimism and more frequently fund direct projects, like land conservation, or reformist political ideas like carbon pricing, instead of investing in a grassroots movement for transformative change.[20]

The fact that most climate communicators prioritize not scaring people should scare you. It means that these communicators *avoid telling the whole truth.* They are not talking about the risk of the collapse of civilization or the deaths of billions of people, even though we are clearly careening toward these catastrophes.

The claims that "fear doesn't work" are not only patronizing and cynical, they have also been devastating in terms of mounting a real and timely response to the crisis. These claims are not supported by evidence. Further, the hollow optimism and positive messaging have tripped the public's bullshit detector. People can tell when they are being given a canned message rather than the unvarnished truth, like at the end of *An Inconvenient Truth,* when Al Gore urges viewers to carpool, check their tire pressure, buy low-wattage light bulbs, and change the settings on home thermostats. We know, with varying degrees of conscious awareness and intellectual understanding, that the Earth's systems are deteriorating more rapidly than these low-effort tips suggest.

The layers of lies, manipulation, and denial add up to a shocking misjudgment of the climate risk by our institutions and government. The "Overton window," also known as the window of political discourse, is a term for the spectrum of policy ideas discussed in the media and considered acceptable. Because public discussion is trapped between denial and gradualism, there is not yet room in the Overton window to accommodate the realities that 1) global warming is occurring more quickly and intensely than reported by the IPCC, and 2) decarbonization is needed in years, not decades. The graph below, from atmospheric and oceanic scientist Michael Tobis, editor-in-chief of the sustainability website Planet3.0, illustrates how the climate conversation fails to consider catastrophic scenarios—even though they are likely.[21]

So just how bad is it? If we don't filter the information to avoid fear? If we don't give into fossil fuel propaganda?

Distribution of professional opinion on anthropogenic climate change

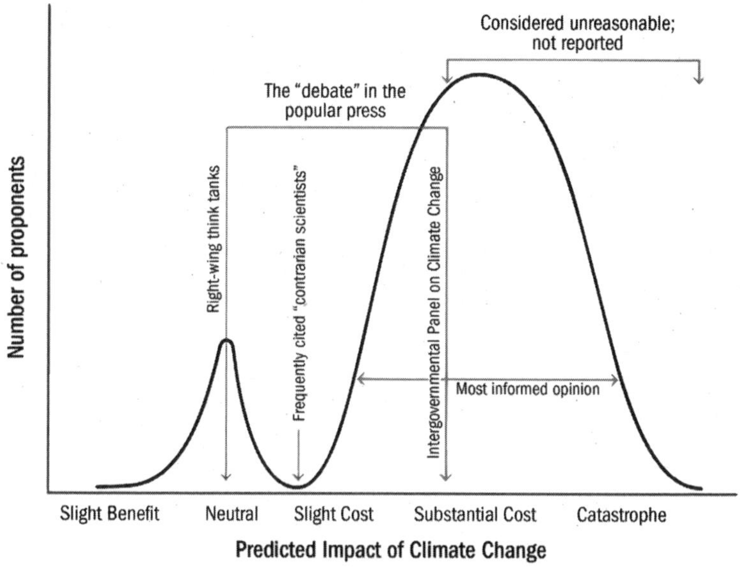

CREDIT: MICHAEL TOBIS AND STEPHEN BAN.

The Earth has already warmed 1.1° Celsius above late-19th-century temperatures. Even this amount of warming is creating devastating droughts and flooding. In fall 2022, flooding in Pakistan displaced 33 million people.[22] Just that one event. The same year, the heat wave in Europe is estimated to have killed more than 15,000 people.[23]

The US media quickly moved on from both. Meanwhile, Hurricane Ian became the costliest storm in Florida history and second only to Katrina for most expensive all-time US storms.[24] Europe, China, Africa, and the United States all experienced severe drought conditions,[25] although "drought" is inexact because it implies a transient period that will end with rainfall; these droughts have continued and will continue to worsen. It's critical to remember that warming is *accelerating*. It is not linear.

This is not the "new normal." This is the beginning of collapse. The widely publicized 2015 Paris Agreement on emissions reductions, if participating countries bothered to follow it, puts us on a path to 2.5°C of warming, as the UN reported in 2022.[26]

We are already well off the map from the climate conditions under which our current civilization developed and grew, in all its complexity. However, 2.5° takes us into a climate in which the earth is unrecognizable. Humans, and the many species we rely on, have never before survived in such conditions.[27]

The gradualist approach of the Paris Agreement and other governments and organizations cannot possibly prevent the collapse of civilization and the living world. Rather, a full-scale mobilization of society for zero CO_2 emissions, plus drawdown, is our last, best hope of canceling the apocalypse and eventually restoring a relatively safe and healthy climate.

Even the IPCC, with its penchant for understatement, recognizes that:

> Pathways limiting global warming to 1.5°C with no or limited overshoot would require rapid and

> far-reaching transitions in energy, land, urban and infrastructure (including transport and buildings), and industrial systems (high confidence). These systems transitions are unprecedented in terms of scale.[28]

Related ecological emergencies such as biodiversity loss, nitrogen and phosphorous flows from fertilizers, ocean acidification, and toxic pollution are exacerbating the emergency of global warming[29] Extinctions are happening at about 1,000 times the rate we would expect without human influence, with dozens of species going extinct every day.[30] The World Wildlife Fund's *Living Planet Report 2018* documents a 60 percent decline in the populations of mammals, fish, birds, reptiles, and amphibians in the last 40 years.[31] The "insect apocalypse" is the tip of the spear, but it's also the base of the food chain. A German study found that the number of insects surveyed by weight decreased by 75 percent between 1989 and 2016.[32] In North America, the Monarch butterfly population has crashed, declining 90 percent in the last 20 years.[33]

As Australian environmentalist and former CEO of Greenpeace Paul Gilding says in *The Great Disruption*, "The Earth is full. Our human society and economy are now so large we have passed the limits of our planet's capacity to support us, and it is overflowing… if we don't transform our society and economy, we risk social and economic collapse and the descent into chaos."[34] In other words, the global economy, human population, and the domesticated animals we raise and consume as food and products have grown so vast and unsustainable that they undermine the life-support systems required for our collective survival. According to the Global Footprint Network, humanity is using the equivalent of 1.7 Earths per year.[35] In 1970, it calculated that we were using the equivalent of one Earth per year.[36]

We are accelerating straight into the collapse of civilization and deaths of potentially billions of people. Our society—our financial, cultural, educational, governmental, and media systems—was built on a safe and stable climate. As crops fail, and hundreds of millions of people are forced to leave their homes, societies become destabilized, states fail and descend into horrific violence. It's already happening: The worst drought in Syrian history caused massive internal migration and unrest, setting the stage for a brutal civil war. The displacement of 33 million Pakistanis is putting tremendous strain on the Pakistani state.

Part of facing climate truth is understanding how the effects of climate change and fossil fuels have been unequally distributed, and how that trajectory will continue unless we build the power to change it. Countries in the Global North have contributed 92 percent of historical emissions, and the wealthiest 1 percent of people create 15 percent of carbon emissions.[37] In the United States, fossil fuel and other toxic projects are almost always located in or near low-income communities of color, who then suffer from air pollution and its health impacts.[38] We have turned communities, and even whole regions and countries, into "sacrifice zones." As food insecurity worsens in the Global South, wealthy people in the Global North tell themselves that climate change is a threat to others, far away.

The UN's *State of Food Security and Nutrition in the World* found that 2.3 billion people were moderately or severely food insecure in 2021, and 11.7 percent of the global population faced food insecurity at severe levels. It also found that 22 percent of children globally have stunted growth, and 6.7 percent of them are experiencing wasting from lack of access to food. In the United States, the richest country in the world, 12 percent of households, and 5 million children, are food insecure.[39]

The UN names climate, conflict, including the invasion of Ukraine, COVID-19, and inflation as the top causes of increased hunger. But, in reality, all of these factors are interrelated, and all point to climate. Crop failure drives hunger, which drives migration and can lead to destabilization and conflict. Russia and Saudi

Arabia would not be able to act as unchecked aggressors if the US, Europe, and other regions didn't depend on them for cheap gas.

How can it be more clear that all of this is a code red for humanity?

These realities have made me view the concept of "progress," as it exists in politics, including "progressive" politics, as antithetical to climate truth. We are not making progress now, and we won't make progress on a rapidly warming planet anytime soon with an extractive and exploitative system. The world is collapsing. Our political vision must recognize that and re-prioritize based on that reality.

I wonder if the climate movement should focus on two core demands: 1) zero emissions and drawdown at emergency speed, and restoration of the biosphere, as these are the building blocks of life and necessary for our collective survival; 2) provision of food and housing for everyone in the world, including climate refugees, as these are the building blocks of a healthy and dignified life.

While feeding and housing all humanity is a measure that wealthy countries should be ashamed of not providing on humanitarian grounds alone—it is also a *civilizational stabilizing mechanism*. As we come ever closer to climate-fueled collapse, quelling people's current hunger and their fear of imminent hunger, as well as rectifying homelessness, will provide immense relief. If we join in a truly collective effort to meet humanity's basic needs for food and housing, we will strengthen our sense of compassion, connection, and moral courage for a collective effort in the race to zero emissions. This will give people all over the world peace of mind and reduce the level of precarity in people's lives, in turn allowing them to more meaningfully take part in our shared project of racing to zero emissions, restoring nature, and healing the world.

This back-to-basics approach is hugely ambitious. Providing food and housing guarantees to all—including climate refugees—would be not only a moral revolution but also a major effort. It would take sustained investment, as well as a transformation in society's outlook on immigration and housing development.

I'm sorry for telling you all of this. I mean it. I wish you didn't have to deal with this devastating situation. I am intimately familiar with the appeal of defending against and blocking this all out. I've already described how, for many years, I did just that, living in the uneasy peace of denial. In high school in Michigan and college in Massachusetts, I was willfully ignorant of global warming and other ecological threats. I was always more interested in people than in nature, more disposed to the humanities and social sciences than STEM. I found discussions of environmentalism and climate change boring. I thought, Why would I care about melting ice, polar bears, or two degrees of warming?

I didn't want to consider that something so seemingly unconnected to me, yet so apparently huge, was something I needed to pay attention to, something that defined my future. So, for years, I utilized the defense of willful ignorance. I've described how I'd start an online article about the climate crisis but get overwhelmed by fear, close the tab, and tell myself, *I can't handle this.* A part of me already knew enough to realize I didn't want to know more.

But over time, as I became more mature and lessened my defenses through psychotherapy, I became increasingly worried about the changing climate. A few events contributed to my transformation. In my senior year at Harvard, I took a course on agriculture to fulfill my science requirement. It opened my eyes to how at-risk the global food system is to climate and ecological threats. After graduating, I backpacked in Europe and Brazil. During my journey, I met many young people from Europe and Latin America who were putting sustainability at the center of their career plans. I saw that many European cities were leagues ahead of the United States regarding mass transit and walkability. I also saw the awe-inducing, acutely threatened Amazon Rainforest.

When I began to discuss my fears in therapy, my therapist thought I was overreacting. She told me, "You worry a lot about the climate, but you don't know much about it." This was a provocative and galvanizing statement for me. I realized that she was right. I had been willfully ignorant for so long that my understanding of the climate crisis was impressionistic and vague.

After my therapist's challenge, I started to finish those online articles. Then, after Sandy, I started reading books and examining scientific studies on the climate and ecological emergencies. The truths I read were shattering. I still remember the force of Bill McKibben's argument in his book, *Eaarth*: We have made such fundamental changes to our planet that the Earth needs a new name.[40] He opens the book with this haunting image:

> Imagine we live on a planet. Not our cozy, taken-for-granted Earth, but a planet, a real one, with melting poles and dying forests and a heaving, corrosive sea, raked by winds, strafed by storms, scorched by heat.[41]

McKibben shows us that we have almost squandered an incredibly rare and precious gift—the Holocene conditions in which civilization developed—and he urges his readers to engage with reality. Reading his book and others like it filled me with terror and heartbreaking despair. I felt, as McKibben advises, as though the alarming statistics associated with the climate emergency were "body blows... mortar barrages... sickening thuds."[42]

This visceral, deeply emotional pain, more than anything else, forced me to take seriously the fact that our civilization and economy are entirely dependent on the atmosphere and biosphere. We are so dependent on the atmosphere and biosphere that they might as well be parts of our bodies. When I started to think like this, to feel the pain of the climate crisis in my own body, I knew my life had to change. If I continued living focused on myself, I was not only shirking my duty as a protector of humanity and all life, I was putting myself at risk. I had to do everything I could to transform our fossil-fueled society and economy.

Once I began to acknowledge the climate's comprehensive impact, I was able to free myself to fully feel the fear and pain that I had been repressing. It felt like the world was collapsing in on me. But it also felt deeply liberating. I was finally confronting the grief, apocalyptic fear, anger, and guilt that I had been working so hard to deny. Rather than relegating them to the corner of my consciousness, where they continued to nag at me, I put those feelings front and center, treating them—and myself—with compassion.

So, I'm also *not* sorry for telling you this. I'm glad that you are reading this and dealing with the painful feelings that it evokes because facing and telling the truth is an incredibly powerful force for change. As you embark on the work ahead, remember that humans evaluate risk socially, not rationally.

Imagine you are at work, and you hear the fire alarm go off. You don't see a fire, and you barely smell a whiff of smoke. How seriously should you take the fire alarm? What if it's just a drill? What if a coworker overheated something in the microwave and triggered the smoke detectors? You wouldn't want to be seen by others as overreacting or becoming hysterical. When you hear an emergency signal but aren't sure if the emergency affects you, you seek answers to questions like these by watching the actions and communications of the people around you, particularly those designated as leaders.[43] If the fire alarm goes off and leaders are chatting and staying put or leisurely exiting the building, you will likely assume that the fire isn't a real danger or that the alarm is just a drill. When people, particularly leaders, react passively to an emergency, so will nearly everyone else. This phenomenon, also known as pluralistic ignorance, has been proven in psychology labs many times.[44, 45]

If, however, when the fire alarm goes off, you see your bosses moving quickly, their faces stern and focused, communicating with urgency and gravity, you will likely assume that there is a real danger and exit the building as quickly as possible.

When we see people acting as though nothing is wrong, it is a cue to us and everyone else that everything is normal. When we see people *disrupting* normalcy and acting as though there is an emergency, we start to view the event as an emergency, too. Telling the truth about the climate, and treating the climate crisis like the emergency it is, is *highly contagious*.[46]

Emergency mode—and what entering emergency mode means for individuals and groups—is discussed further in Step Four. Here, I just want to point out that Americans as a nation have entered emergency mode before.

Most famously, Americans joined the Allies in emergency mode when we mobilized to win World War II. Faced with the prospect of annihilation, all Americans were expected to pull together by working in war jobs, growing victory gardens, contributing to scrap drives.[47] The US government made an extraordinary investment in the war effort, spending more than 40 percent of its gross domestic product on the war effort,[48] issuing strong regulations, and rationing critical goods like gasoline and meat.[49]

The home-front mobilization during WWII is the example par excellence of how an entire society can enter emergency mode. During WWII, our whole country entered emergency mode. What the United States, the United Kingdom, the Soviet Union, Canada, and other allies accomplished during the 1940s is not only inspiring, it's a model for how we can collectively and effectively enter emergency mode today. We must look to our country's collective response to the threat of Axis power, to the specific methods by which we transformed our economy, shattered every production record to arm ourselves and our allies, and won WWII.

After years of bitter acrimony over the New Deal, WWII prompted conservative business titans and "New Dealer" government officials to join together, albeit in an uneasy alliance, to focus America's industrial might against the Nazis and imperial

Japan. Factories were rapidly converted from producing consumer goods to producing tanks, guns, bombs, and planes. The quick turnaround and staggering output shattered all historical records for war production.[50]

During this time, all hands were on deck, and everyone contributed what they could: Young men sacrificed their lives fighting for our country, women surged into factories to produce war material, Native American Code Talkers developed a system based on several Indigenous languages to transmit secret messages for the allies,[51] and scientists and universities pumped out research on behalf of the war effort, leading to huge technological and intellectual breakthroughs. Fifteen million Americans, more than 10 percent of the population, relocated to find a war job, often across state lines,[52] and more than 40 percent of vegetables were grown at home in Victory gardens.[53]

The response to WWII led to what I call emergency mobilization. This is when society enters emergency mode, responding to an emergency by collectively directing its energies to immediately restructure its industrial economy and society. Emergency mobilization requires the support and involvement of the vast majority of citizens and the redirection of a very high proportion of available resources, and it impacts every part of society. Emergency mobilization is nothing less than a government-coordinated social and industrial revolution.

When an entire nation enters emergency mode, the results can be truly staggering. By mobilizing for total victory, the United States achieved goals it could never have reached in another way. When the United States entered WWII after the attacks on Pearl Harbor in December 1941, President Franklin D. Roosevelt (FDR) laid out terrifically ambitious production targets for tanks, ships, guns, and airplanes. FDR set the goal of producing 60,000 planes in two years. People were deeply skeptical about whether such a feat could be accomplished. Yet, by 1944, the United States had produced 229,600 planes—more than three times the original goal.[54] In response to a cutoff of critical rubber supplies in Southeast Asia, the federal government launched a program that scaled up synthetic rubber production from less than 1 percent to about 70 percent of total US production—a *100-fold* increase—in about four years.[55] In 1943, reclaimed rubber from citizen-coordinated scrap drives provided about 40 percent of domestic rubber production.[56]

America also made huge scientific advances, in part by generously funding research of many kinds.[57] The first computer was invented during this time, as were plasma transfusions and sonar technology. The Manhattan Project successfully built the world's first atomic bomb in less than three years—a morally fraught but stupendous feat of planning, cooperation, and scientific ingenuity. As an integral part of the mobilization during the multiyear emergency of WWII, the United States managed to maintain and, in some cases, expand its basic systems, including infrastructure, education, health care, and childcare, and ensure that the basic needs of the civilian economy were met.

We do not need to have an overly rosy view of that time to appreciate the transformative effects of the United States' commitment. We cannot overlook the racist policies and attitudes—the military and many industrial mobilization jobs were segregated, and more than 100,000 Japanese Americans were interned. However, we can accept the painful reality of the United States history of racism and still acknowledge the transformative potential of mobilization.

In fact, during WWII mobilization, major strides were made toward both racial and gender equality. FDR created the Fair Employment Practice Committee to investigate discrimination claims in response to the fierce organizing of civil rights leader A. Philip Randolph and the March on Washington Movement.[58] The Double V Campaign advocated for victory for democracy and equality for African Americans overseas and at home.[59] Five million women joined the workforce for the first time,[60] and daycare centers were built in factories to provide childcare support.[61] Some factories also provided mothers with prepared dinners so that they could work a full shift and still provide a hot meal for their children.[62]

A sense of national purpose and incredible energy suffused the entire country.[63] Because they were in emergency mode, citizens felt intensely motivated and made many sacrifices. They invested their cash savings in war bonds.[64] They tolerated a significant increase in taxes: The percentage of the population paying any income taxes jumped from 7 percent to 64 percent. Tax increases focused on the rich. The top marginal income tax rate on the highest earners reached 88 percent in 1942 and a record 94 percent in 1944 on income above $200,000—the equivalent of about $2.85 million in today's dollars.[65] Further, taxes on excess corporate profits were implemented, with rates as high as 95 percent for the most profitable corporations.[66]

The federal government instituted a sweeping rationing program to ensure the fair distribution of scarce resources on the home front and to share the sacrifices equitably. Gasoline, coffee, butter, tires, fuel oil, shoes, meat, cheese, and sugar were rationed. American historian Doris Kearns Goodwin describes the impact of these measures:

> By and large, American housewives accepted the system of rationing cheerfully.... Citizens learned to walk again. In the following months, car pools multiplied, milk deliveries were cut to every other day, and auto deaths fell dramatically. Parties at homes and nightclubs generally broke up before midnight so that people could catch the last bus home. All in all, pleasures became simpler and plainer as people spent more time going to the movies, entertaining at home, playing cards, doing crossword puzzles, talking with friends, and reading.[67]

We also entered something like emergency mode during COVID-19, although this was a case study with a more mixed approach and therefore with more mixed results. In March 2020, in just a few weeks, a shift happened all over the world. We saw how rapidly governments and individuals can change their behavior. Governments around the world sent cash to people, creating a temporary universal basic income program. Operation Warp Speed led to the rapid development of MRNA vaccines. People transformed their lives, their businesses, and their plans in response to COVID-19. Americans stayed home, wore masks, and postponed birthdays, graduations, and other celebrations.

In the United States, low levels of social trust, the lack of public health infrastructure, the failure of leadership from the CDC, and the cynical exploitation of the crisis by the Trump administration and its allies held us back from fully entering emergency mode and undertaking transformative projects, such as installing high-grade ventilation systems in buildings and air-quality monitors in every public space.

Even so, CO_2 emissions fell by 2.4 percent in 2020,[68] which is approximately the pace needed to reach the (inadequate) Paris targets. While not as resounding, or lasting, a success as the home-front mobilization during WWII, 2020 does provide an example of a time when almost every human being on Earth changed their behaviors rapidly to meet a threat.

Today, even though it's our entire planet that's burning, there hasn't yet been anything like this level of emergency response to the climate crisis. Our collective denial, fueled by our aversion to feeling fear and grief, holds us back. By denying the truth of the climate emergency, we remain in the thrall of helplessness, our political system remains intractable, and our politicians and mainstream climate movement organizations remain unwilling or unable to lead the way. Virtually everyone—even powerful leaders—appears overwhelmed and incapable of processing the true scale of the climate crisis. We all feel helpless, unable to grasp how we can meaningfully contribute to transforming our world and protecting humanity. We look to our leaders for our cues, but we only see the same passivity, powerlessness, and helplessness reflected back to us.

This sense of powerlessness is a myth. It's rooted in a lack of understanding of how malleable the human condition is, how capable of growth we are, and how people have formed social movements and created fundamental changes in society over and over again. The feeling of powerlessness has been fostered by neoliberal ideology. Instead of viewing ourselves as citizens in a democracy—as people who are responsible for creating the world we want to live in together—we have been taught to view ourselves as a group of self-interested individuals whose only role is to compete in the free market and "get ours." This ideology has dominated the American and global political imagination for decades. It tells us that our society is a meritocracy that rewards hard work and virtue, it tells us that poverty is a moral failing, it tells us that there is no alternative to our current economic model and that only "market-based" remedies for the climate are "realistic." We've been told, most famously by former British Prime Minister Margaret Thatcher, that "there is no such thing as society," and certainly no group of people joined together in the common interest of a habitable planet.[69]

However, I'm here to remind you that groups of concerned citizens have changed the world many times before—and they have done it through the power of truth. Leveraging the power of the truth has been a central strategy in successful social movements. Such movements do not discover brand new truths. Instead, they manifest truths that are widely known but have been repressed, denied, and ignored by governmental, corporate, media, academic, and religious institutions.

We've seen this dynamic recently in the racial justice uprising of 2020, and the #MeToo movement in 2017, but it has happened throughout history.

For example, before the Civil Rights movement brought the ugly truth of racism to the forefront of American life, most white Americans ignored or passively accepted Jim Crow, telling themselves that it was not their problem. Yet, when nonviolent protesters were met with hateful violence and those events were broadcast on television across America, the truth could no longer be denied or ignored. In seeing the clear truth of racial segregation's moral bankruptcy, viewers across America could no longer claim that this didn't concern them. People were forced to choose a side; when more Americans began to take the side of civil rights, seeing it as their issue, major immediate changes were seen as undeniably necessary. When a powerful truth, such as the racist, brutal treatment of African Americans, is communicated widely and effectively, transformation can happen rapidly.

The Velvet Revolution offers another example. The desire for change in Czechoslovakia had been building throughout the 1970s and 80s. Czechoslovakians wanted freedom from the repressive USSR government. Artist and dissident Václav Havel led a successful social movement dedicated to spreading that basic truth. Havel, who championed "living in truth" rather than complying with the corrupt, repressive actions of the Soviet state, argued that the government's lies would collapse when the force of the truth was unleashed. He was right.

When Havel began his work, dissent was illegal. Most people were either afraid to speak up or thought it was pointless because others, they thought, supported the regime. In an enforced state of pluralistic ignorance, people underestimated the popularity of their own cause. Dissidents like Havel worked to contradict and expose this pluralistic ignorance, demonstrating that we are not alone, we all want democracy. The illusion was shattered on November 20, 1989, when 200,000 people demonstrated in Prague. When others realized how much power was on the people's side, the number of demonstrators swelled the very next day, to 500,000 people, and a general strike was called. Ultimately, the entire leadership of the

Communist Party of Czechoslovakia resigned, the first democratic elections in forty-three years were held, and Havel was elected president.[70] This collective upheaval was possible because, as Havel asserted, the truth has inherent power:

> [The power of truth] does not reside in the strength of definable political or social groups, but chiefly in a potential, which is hidden throughout the whole of society, including the official power structures of that society. Therefore this power does not rely on soldiers of its own, but on soldiers of the enemy as it were—that is to say, on everyone who is living within the lie and who may be struck at any moment (in theory, at least) by the force of truth (or who, out of an instinctive desire to protect their position, may at least adapt to that force). It is a bacteriological weapon, so to speak, utilized when conditions are ripe by a single civilian to disarm an entire division.... This, too, is why the regime prosecutes, almost as a reflex action, preventatively, even modest attempts to live in truth.[71]

Although the truth is a radical and incredibly motivating force, it must be made manifest to have power. When it comes to the climate crisis, it's not that the people are powerless; it's that larger forces have conspired, both purposefully and inadvertently, to deny the truth, making us feel helpless. Most of what we've been told about the climate emergency is either outright deceit (on the part of the fossil fuel industry and the GOP) or euphemistic understatement (on the part of the Democratic Party and the gradualist climate movement). Is our inability to effectively respond surprising? Is it any wonder that so many of us do not—or are simply unable to—acknowledge and respond to the climate crisis as the emergency that it so clearly is?

Thankfully there is a robust movement that is committed to telling the truth, disrupting normalcy, and building power. I call it the climate *emergency* movement. This movement demands what science and morality tell us are necessary—absolutely no more expansion of fossil fuel infrastructure, instead we need a race to zero emissions that takes ten years or less, plus drawdown and massive ecological restoration. The climate emergency movement is dedicated to disrupting normalcy because normalcy threatens us all.

Questions for Reflection and Discussion

- How did it feel to read this Step? What information stood out to you?
- How well do you understand the scale of the climate and ecological crises? Do you feel you need to learn more to engage effectively?
- Have you found facing the truth to be transformative in other areas of your life?
- When have you experienced personal growth by facing a hard truth?
- What defenses have you used to protect yourself from climate truth? Do you think you will be able to lessen or let go of those defenses?

STEP TWO:
Welcome Fear, Grief, and Other Painful Feelings

Only when we are brave enough to explore the darkness will we discover the infinite power of our light.

—Brené Brown

FACING THE CLIMATE EMERGENCY is hard. It challenges us to reach new levels of emotional strength, maturity, and wisdom. That's why a critical part of facing the truth means improving how you relate to your feelings. Rather than block out or deny these feelings, you must face and work through your fear, grief, anger, guilt, and all the other emotions you've repressed about the climate. In fact, you must go further than just facing them—you must learn to welcome and even honor these painful feelings. Only then will you take back the control these feelings have over you. Only then will you shake off their numbing and paralytic effect and be able to use their power to transform yourself and our society.

In this Step, I will help you notice your pain with nonjudgmental interest, welcome your pain, and turn it into fuel to propel you toward heroism.

First, you must learn to feel your feelings. Easy enough, right? Nope. It's one of the hardest things for humans to do, especially in an alienated society like ours. Growing up, we are taught that some feelings are not acceptable and must be repressed and denied at all costs. For most people, the emotions that were scorned, feared, and otherwise rejected in our families of origin are feelings that we've built a lifetime of defenses against. Often, these lessons take gendered lines: Boys are taught that fear, vulnerability, and sadness are pathetic and embarrassing, while girls are taught that anger makes them "bad girls."

Too often, when we look inside ourselves and find "negative" feelings, we usually judge and punish ourselves:

- I can't feel rage and hatred toward my mother. I owe her everything.
- I can't ever feel overwhelmed by grief and despair. That would be pathetic and humiliating.
- I can't feel sexual attraction toward that person. That would be totally perverse, inappropriate, and just wrong!
- I would never think something sexist or racist. What am I, a monster?

When we feel the feelings we've built our defenses against, we typically respond with intense self-judgment, telling ourselves to "get over it" or that we're a horrible person. Yet all feelings are a normal, and even essential, part of the human condition. Censoring and judging thoughts and feelings usually makes us feel worse—and it certainly doesn't make them go away. Quite the opposite. If we can't acknowledge our feelings, they maintain tremendous power over us. Psychotherapists and meditation teachers understand that we are healthiest when we accept—and then allow ourselves to experience—all of our thoughts and feelings without judgment and with compassion.

The best approach in almost any situation—even the most painful—is to nonjudgmentally recognize what we are feeling, consider the situation—including how our values should inform us—and then act based on a synthesis of our feelings and more rational evaluation. When we deny our emotions, we can't reach this synthesis, and we stay stuck in the feelings we want to deny. This "stuckness" is the ironic reason that people who deny their feelings often end up being the most dominated by their emotions.

Psychotherapists also know that you can't selectively filter out certain feelings without generalized consequences. If you are blocking anger and grief, for example, you will also constrict your ability to experience love and joy. It's not the content of our negative and painful thoughts that matter—what matters is how we face and process those thoughts and feelings and, ultimately, how we act. You've probably already noticed that your negative and painful feelings exert power no matter how much you avoid, deny, or tell yourself to "get over" them. But they do something else, too. When, for example, you tell yourself that you're a terrible person because you had a racist or sexist thought, you fail to explore, understand, and challenge your prejudiced ideas, inadvertently short-circuiting what could have been a self-reflective change process. Welcoming your feelings is a prerequisite for real growth.

Although humans are masters at avoiding painful and uncomfortable feelings, we can learn to face and accept them by practicing self-compassion and building what psychologists Kerry Kelly and Jack Novick call "emotional muscle." Building emotional muscle is core to mental health.[1] When you process your feelings in a healthy way —facing, accepting, and investigating their presence instead of attempting the fruitless work of repression—you accept your whole self. Most importantly, expanding your emotional range allows you to live in reality and respond to it in a moral and effective way. If we can't feel our feelings, we can't face reality, and we can't respond to it with our actions.

To welcome your feelings and build emotional muscle, approach your pain with an attitude of curiosity and self-compassion. Easier said than done, but you probably already have a powerful template for practicing such self-compassion. According to Kristin Neff, associate professor of Human Development and Culture at the University of Texas at Austin, offering ourselves compassion means offering the same nonjudgmental comfort we offer our close friends.[2] To build your emotional muscle, practice treating yourself and all of your feelings just the way you would treat a beloved friend who came to you for help. You wouldn't tell them to ignore their pain or call them a bad person. You would instead realize that they needed your help and—if you're a good friend—you would listen to their feelings with interest and respond to them with compassion and empathy.

In addition to our efforts to avoid, deny, and repress to avoid feeling our feelings, we also dissociate. Dissociation disturbs the normal links between thoughts, feelings, and actions. It can range from mild "zoning out" to out-of-body experiences, amnesia, or, in extreme cases, fugue states or multiple personality disorder. Dissociation is a last-ditch defense that occurs when a mind is overwhelmed; it is frequently experienced during traumatic situations. Dissociation protects us in moments when nothing else does—this is why children often dissociate when they are exposed to violence. But dissociation isn't a healthy long-term strategy. When you dissociate, you lose touch with what is happening around you and what you know is true.

When we dissociate from the pain of the climate crisis, we protect ourselves from feeling the discomfort of pain, but we do not protect ourselves from catastrophic climate breakdown. Instead, when we dissociate to manage our overwhelming feelings, we become stuck at the level of apathy and willful ignorance. We may protect our feelings, but we do so at the expense of our actual physical safety.

As you undertake the work of facing climate truth and begin to work toward transformation, you may feel like you're constantly fighting the desire to dissociate—to watch TV, to zone out, to get drunk, or to self-medicate in some other way. It is true: Facing the reality of danger and destruction on this planet is terribly painful. But allowing and accepting this pain is necessary for the work ahead. If you seek to move forward in reality, to help restore a safe climate and protect humanity and the living world, you must allow yourself to experience the grief, horror, terror, guilt, anger, and other feelings that the truth evokes.

Any meaningful, world-transforming social movement consists of thousands of individuals who have already changed their outlook and their behavior and are leading the way in creating a better world. These individuals are living in climate truth.

Living in climate truth is hard, but it's the only way forward. Once you begin to live in climate truth, you will experience a wide range of emotions, and they will be intense. They will be upsetting and overwhelming, but that is wholly appropriate to the stakes of a crisis that is set to lead to mass destruction and death. Your painful feelings spring from the best parts of yourself, from your empathy, sense of responsibility, love for others, and love of life. These feelings connect you to all life and will fuel the work ahead. Immersing yourself fully in them is a heroic, even sacred, undertaking.

The work you do in learning to accept all feelings as they arise will enable you to accept, to feel, and to use the intense emotional reactions that will result from living your life in climate truth. When you accept and process these painful feelings, you are also better able to access, accept, and process the pain of others. The more comfortable and confident you are with fear and pain, the more you will be able to help others accept their own intense feelings and turn them into action. This, too, will be necessary because, as we must strive to remember, we are all on this rollercoaster together.

Psychoanalysis provides critical insights and tools regarding how to face climate truth, emotionally as well as intellectually. I believe that psychoanalysis and the various talk-based therapies it launched are some of the great breakthroughs of the twentieth century, on par with computers and antibiotics. I wish everyone in the world could access high-quality, emotionally supportive therapy. It can help virtually every problem, if just by providing a place to practice and experience empathic listening. For me, therapy has helped in every area—and almost every stage—of my life. In my darkest times, therapy kept me afloat; in better times, as I sought to transform myself in service of the mission, it supported me in reaching higher levels of insight, affect tolerance, acceptance, and self-compassion. Maybe psychotherapy can help you as much as it helped me. Even if you've never considered therapy, even if you think you rarely practice avoidance or dissociation and seldom censure your self-talk, psychotherapy can help you explore yourself and grow.

You'll be in good company. Some outstanding social movement leaders have found help in psychotherapy. For example, AIDS activist Larry Kramer, discussed in Step Four, said that his years of psychoanalysis during college helped him confront and channel his anger, fear, and grief into effectively responding to the AIDS crisis when others were overwhelmed.[3] Pope Francis, a strong moral leader on the climate crisis, also sought the help of talk therapy in his forties.

Think of therapy as hiring a personal trainer—one who helps prepare you for the marathon of life. This is particularly true for your life as a climate activist. I know it's not always possible to consult or hire a professional. I am privileged to have been able to afford and have the time required for therapy. Readers who are prohibited by cost but not by time and are interested in pursuing psychotherapy might consider a lower-cost option at a psychoanalytic institute. Such institutes are staffed by advanced trainees, psychologists, psychiatrists, and social workers who are becoming psychoanalysts. These psychoanalytic trainees are often willing to take on clients at a reduced rate.

If that's not an option, you can still reap some of the benefits of these methods without a therapist. You can, for example, practice responding to your thoughts and feelings with curiosity rather than judgment. As described above, in this practice, you work hard to stay with your feelings. You don't downplay or ignore them or attempt to intellectualize them. Instead, when you feel uncomfortable feelings, you assume an attitude of active interest and self-compassion.

To offer a sense of how this might work, let's try it out. Let's say that reading this book makes you feel overwhelmed, helpless, and cynical. Maybe you're so upset by my words that you're contemplating throwing this book out the window. Perhaps you're starting to think some negative, angry things about me.

Despite the naturalness of this kind of reaction, most readers have trouble responding to negative thoughts with understanding. For many readers, self-judgment kicks in harshly. They register their negative feelings toward this book or me and tell themselves they're pathetic. They may berate themselves, saying, *You say you care about the environment. How can you give up like this? Are you weak? A coward?* Other readers work hard to shut down their negative feelings. Still other readers will argue with themselves, questioning their intelligence or authority, or blame themselves for their feelings.

But what happens when, rather than trying to prevent the negative feelings you may have as you read this book, you assume an attitude of active interest and nonjudgmental curiosity? They are, after all, just feelings. They don't hurt anyone or impact the world—only your actions do. Take a few deep breaths. As you become aware of negative feelings, such as a desire to deny or an urge to defend, or a need to express skepticism, don't try to talk yourself out of these feelings. Instead of judging your feelings—and thus yourself—try to be curious about them, allowing yourself to really and fully experience them. Try, for example, to name the feeling—not justify, evaluate, or rationalize, just name: This book is making me feel angry. Notice where in your body your anger is being experienced. Are you holding tension in your shoulders? Are you clenching your teeth?

By exploring your feelings, you become more comfortable with them, and you build your emotional muscle, increasing your ability to face other, more painful emotions, such as those you will feel when you fully face the climate emergency and follow the call to protect humanity.

There are other approaches to learning self-compassion, increasing your affect tolerance, and building emotional muscle.

For example, you can identify the friends or family who are most comfortable talking about their feelings and who make you feel comfortable and accepted when you talk about yours. Make a regular effort to seek out these people to talk with about your feelings and theirs. Most of us, especially those not in therapy, are not in the practice of talking about how events and people make us feel. We live in a culture that frequently rewards surface pleasantries on the one hand and smart judgments on the other.

But when you practice talking about your feelings, you get more comfortable noticing, identifying, and tolerating them. Start with relatively easy feelings: *I felt hurt and disappointed when my friend canceled the party, or I feel worried and anxious about my upcoming test.* Getting comfortable identifying and sharing these kinds of feelings and having them received nonjudgmentally prepares you to share other, more complex and challenging feelings.

Mindfulness and meditation offer another approach. Clinical psychologist and Buddhist teacher Tara Brach discusses the benefits of meditation, using the acronym RAIN as a guide:[4]

- R: *Recognizing* what is happening.
- A: *Allowing* life to be just what it is.
- I: *Investigating* inner experience with gentle attention.
- N: *Nurturing* ourselves in our experiences.

RAIN helps its practitioners stop seeing themselves as being the same as their feelings. Our feelings do not define us; they happen inside us, and we can nonjudgmentally notice them and choose how we react to them.

There are many other tools and practices that can prepare you to face your feelings and ready you to transform through climate truth. You might, for example, consider keeping a journal or log of your thoughts and feelings. For many people, this work can be another way to gain access to RAIN-based insights. By putting your feelings on paper, you create the distance that can help you accept them, no matter what they are, with self-compassion.

You'll know you are making progress when you allow yourself to experience more thoughts and feelings without judgment and censorship. You should feel proud when you notice yourself experiencing feelings that you are uncomfortable with—especially those that make you feel guilty or ashamed. If you have trouble crying, honor and praise yourself when you do cry. It is truly a victory. Every time you allow yourself to feel hard feelings, you expand your ability to tolerate affect. You build your emotional muscle. You make yourself stronger.

What else can help? Well, I got two pandemic puppies, Hero and Cassandra, that bring me daily joy. In addition to learning to cope with your emotions in therapy, facing climate truth for the long haul requires authentic self-care—not the temporary solace that comes from buying things, not numbing out from substances, but the actions that bring us daily joy and restoration, like exercise, regular sleep, healthy meals, and spending time with loved ones. Including pets. We all deserve to experience joy even in these dire circumstances.

For me, it was only when I began to study and understand the climate's comprehensive and devastating impact that I was able to free myself from the nagging, shadowy fear that had been haunting me by fully feeling the fear and pain that I had been repressing. When I began to face climate truth, I felt like the world was collapsing in on me. But I also felt deeply liberated. I was finally confronting the grief, apocalyptic fear, anger, and guilt that I had been working so hard to deny. Rather than relegating them to the corner of my consciousness, where they continued to haunt me, I put those feelings front and center, treating them—and myself—with compassion.

I share this with you so that you might recognize similar feelings and experiences and treat yourselves with compassion in feeling them. We are, after all, two people trying to make sense of this horrifying reality we find ourselves in. At this late hour, reckoning with ecological reality is at the core of what it means to be truly engaged with life.

My fear is a constant presence in my chest and stomach. I am afraid of the unbearable pain I will feel as civilization continues to collapse, as communities, institutions, and states fail, as more and more people are immiserated, and as species become extinct.

Fear has become my greatest motivator. It helps me keep other motives, such as my desire for narcissistic gratification, in check. I am a competitive person, so concerns about "getting credit," or being the "best," or "directing the most popular" organization nag me. However, I don't deny my feelings when I feel envy or competitiveness toward another organization or climate activist. Instead, I acknowledge them and then ask myself an honest question: *Do I want to be a big shot, or do I want to prevent collapse? Which is my most important priority?* My answer is always the same, and I am made to remember how small and irrational my narcissistic ego needs are. Note that I do not judge myself harshly for feeling competitive or ego-driven. It's fine. It's human. I simply remind myself that those feelings are not in line with my values and priorities, and I refuse to be driven by them. If I denied my competitive feelings or judged myself for experiencing them, I would be much more likely to act from feelings of competition and judgment.

In addition to fear, I feel a seemingly bottomless sadness. I am heartsick about the ecological crisis. There is so much suffering in the world now, and we are heading straight into total devastation. Human life and our natural world are the greatest blessings imaginable. This is true whether we believe life was given by God, by other spiritual forces, or by randomness. The most intelligent species is destroying itself and bringing on a sixth mass extinction. We are destroying our greatest gifts. We are choosing death.

I cannot be cynical in the face of all this loss and suffering, telling myself that humanity is irredeemable and collapse is inevitable. For me, this tragedy is all the more painful because I believe in the glorious immensity of human potential. I know that growth and change are possible in individuals and societies. In fact, as a psychologist, I believe that the tools that I've been talking about—processing and accepting feelings with nonjudgmental self-compassion—are crucial to achieving growth and transformation. Humans are excellent at responding collaboratively and effectively when facing existential crises. That is why the gap between who we are and what we're doing now and what we could be and do is so devastating. We are capable of so much more than this.

I also feel disgust, shame, and contempt. I am disgusted by the dehumanizing and racist death machine we call an economic and political system. I am disgusted and ashamed of myself for taking part in it. I feel disgust and contempt for everyone else who takes part in it. I feel filthy for taking part in it—it's a stain I cannot wash off. Every day, we pump more carbon into the atmosphere, put more plastic into the ocean, cause the extinction of more species, and we do it with a smile! (Fear doesn't work as a motivator, remember?) I sometimes wonder if, when we arrive at the pearly gates, and St. Peter is determining whether to let us into heaven, we will be faced with a pile of garbage and a $C0_2$ calculation: Time to tally up all the plastic crap and everything else we ever threw "away."

I feel incandescent rage. It's always there, coursing through my veins. What we are doing is wrong. It's evil. I want to shout and scream. I have had confrontations with drivers idling their cars on the streets of New York City, but I try to avoid them. I feel angry when people post on social media about their tropical vacations. I feel angry about people "just living their lives" as normal.

I am sickened by the Democratic Party and the progressive movement for treating the climate crisis as a side issue for so many years. I am angry that the environmental movement hasn't had more success, and I am full of critiques about their focus and goals. I am angry at Baby Boomers—my parents' generation—for allowing our crisis to reach this point.

Although some people feel visceral anger at oil company executives or GOP politicians, I guess I expect that evil people do evil things. I agree they have committed crimes against humanity and should be tried at the International Court of Justice. I feel angrier with people I know—and often people I love—for failing to protect me and all life.

I feel betrayed by my family members who voted for Trump. But also betrayed and abandoned by family members who support my climate activism as "my thing," but don't recognize that it needs to be "their thing," too.

I sometimes feel destructive glee. I feel so rageful about collective denial—and so disgusted by our death-machine culture and economy—that I think something like, *Don't come crying to me when the crisis is at your door. You had your chance; I warned you about this.* It's kind of a ghoulish self-righteous revenge fantasy. In this feeling, being "right" becomes more important than protecting humanity and the living world.

I feel guilty for not doing more to solve the climate crisis: not working harder, not being more effective, not personally getting arrested, not donating more money, or otherwise sacrificing more. I feel guilty about my consumption, as well as all of the horrible things happening in the world, all of the present suffering and oppression that I am not focused on because I am focused on preventing a catastrophic breakdown in the near future. I feel guilty for my many privileges, my life of comfort. And why not? The world is so brutal and unfair. How did I end up so lucky? Sometimes I want to renounce this fallen world and live like a monk or a nun.

I feel morally responsible for doing everything I can to prevent the full unfolding of the climate crisis. I wake up feeling the heaviness of this responsibility. It stays with me all day. It pushes me, it eats at me. Although I can sometimes relax and socialize, those pursuits are secondary to my true mission. I sometimes feel tremendous pride in Climate Emergency Fund's accomplishments, and the accomplishments of the activists we support, but until the American economy is fully mobilized and we are racing to restore a safe climate, I do not feel that I've met my responsibility.

I also feel despair and helplessness. I am so small and the denial machine is so large. Sometimes I am without hope. I think that my efforts, and the efforts of many other passionate and dedicated individuals, are doomed to fail. It is too late; humanity is irredeemable. There is nothing I or anyone else can do.

Sometimes, I want to die. My despair, terror, and guilt about the collapse of civilization and mass extinction of species have made death seem like sweet relief. For some readers, this admission might sound shocking, but most people I know who live in climate truth grapple with this feeling at times. Frankly, it's an understandable response to the crisis. We are living in a broken world, in an age of mass violence and death. The crisis is global, and there is no safe haven. Death can sometimes feel like the only way to end the nearly unbearable feelings and escape the hellish future we are careening toward. However, in this instance, it is especially critical to distinguish between thoughts, feelings, and actions. As with all feelings, it is healthier to nonjudgmentally acknowledge the feeling of wanting to die than to attempt to deny it.

I want to be really clear: No one should kill themselves in response to the climate emergency. If you are feeling suicidal, put down this book right now and call 988, the national mental health hotline, or for those outside the US who need support google "suicide hotline" and the name of your country to find more localized support.

It's not time to give up. Not while there is still any chance of restoring a safe climate. We should choose the more challenging path of going all-in to solve the climate crisis. Let the pain be your fuel. Let your total rejection of the status quo give you the courage to transform your life, to stand out from the crowd, and to demand transformative action.

I feel alienation. I feel so different from people who aren't living in climate truth. When I walk down the street in Brooklyn, and everyone is going blithely about their lives, I feel strange and different from everyone. I feel they cannot really understand me and that I cannot understand them and what I see as their petty or self-involved concerns.

Of the range of emotions that I describe, two deserve special consideration: fear and grief. In the prior Step, I discussed how the institutional climate movement gives fear a bad rep. In truth, fear and grief are *especially* worth honoring, as they are particularly potent in transforming yourself and society, and avoiding them has been especially destructive to the climate movement's efficacy.

Fear is one of the seven primal emotions that mammals experience and one of the most reliably effective motivators for the entire animal kingdom. It evolved in animals to force a response to threats.[5] Fear helps us protect ourselves; it mediates between perceiving danger and taking defensive action. Fear triggers us; launches us into action.[6]

That's why, among the many and varied emotions we are suppressing about the climate crisis, fear is the emotion we need to feel the most. Fear is the mechanism through which we turn the perception of danger into self-protective action—without it, we expose ourselves to terrible risks.

We feel afraid because we know that we are not separate from the natural world but that our lives—now and forever—depend on a steady climate and biosphere. If the oceans die, we die. If the forests die, we die. It is the most noble parts of ourselves that want to protect each other, and it is a testament to how much we cherish our own lives, our families, our communities, the abundant species with whom we share this Earth. In this way, fear is a beautiful feeling, in part because it is a protective feeling. Listen to it urging you to protect yourself and all life!

If you have not experienced that fear yet, consider that you might be blocking or avoiding it; try to make space for it and invite it in. Try to visualize what the collapse of civilization will look like and imagine what it will feel like. Use your imagination to explore what it will feel like: No water coming from the tap; not knowing whether to stay or go; mass starvations and displacements everywhere. If you can't imagine it for yourself, read Octavia Butler's *Parable of the Sower,* a novel set in 2025 in partially collapsed Southern California. Butler's depictions of mass migration—thousands of people walking north on the LA freeways amidst profound social breakdown—may be the jolt you need.

Feeling your fear and grief will hurt. But don't worry, the pain is there for a reason. Also, it's not a new pain. It has been with you your whole life—no one living on this planet can avoid the ecological crisis entirely. Practicing the self-compassion that will allow you to feel and keep feeling the fear and its resulting pain will propel you forward into committed action; it will help you go all-in.

When I began to learn more about the climate emergency, I felt despair. I knew I needed to grieve and mourn. I knew that the grieving process was the only psychologically healthy response to incalculable loss. It's seldom discussed outside a therapist's office, but grief is the best method we have for reacting to losses and adapting to new realities.

Therefore, once we face the truth and feel our feelings associated with the climate emergency, we must allow ourselves to experience and work through our grief. Grief is painful, but if we stop ourselves from feeling it, we stop ourselves from emotionally processing the reality of our loss. If we can't process reality, we can't live in reality: We become imprisoned and immobile.

Grief ensures we don't get stuck in denial, living in the past or in fantasy versions of the present and future. Think of the widower who cannot acknowledge the death of his wife, who cannot cry, and who never cleans out her closet. He remains suspended between the past and his fantasy version of the present. Because he can't grieve, he can't adapt to his new reality and cannot find any satisfaction in what is a new—if unwelcome—stage of life.

According to Joanna Macy, an environmental advocate and spiritual teacher, we must grieve to accept the reality and pain of loss. In her book, *Active Hope: How to Face the Mess We Are in Without Going Crazy*, Macy writes that grief follows from a heart that has been broken by loss. The initial task of our grief is simply to help us accept the actuality of the loss, to face the truth. The second task of our grief is to allow ourselves to suffer, to feel the agony of what we have lost, which helps us understand the meaning and depth of our loss:

> When we feel this emotion [of grief], we know not only that the loss is real but also that it matters to us. That's the digestion phase—where the awareness sinks to a deeper place within us so that we take in what it means.[7]

Macy argues that we acknowledge reality through grief, that grief helps us to "find a way forward that is based on an accurate perception of reality."[8] When we refuse to grieve what we have lost, we lose the opportunity to celebrate and honor what we love.

I went through several months of intense grief as I was facing climate truth, and I continue to grieve. I have cried many times; I cried about my fears for myself and my family, the starving people in the Global South and right here at home, for the gorgeous species that are going extinct. Feeling grief reinforced my recognition of the seriousness of the climate emergency and the seriousness of our collective loss. But grief did something else, too. It reminded me of how much I love this world. The depth of my grief was a direct response to how connected I am—and want to be—to the living world. It gave me a new breath of life.

This makes grief not only worth experiencing but worth honoring. We've lost so much. Millions of people have already died because of the climate and ecological crisis, mainly because of hunger and infectious diseases.[9] Most are among the world's poorest people. We grieve them because they matter.

It's not just humanity that has been lost. Biodiversity—the riot of life—is also being speedily destroyed. We have already seen thousands of species slip into extinction. The species that still exist are rapidly losing numbers. Nisha Gaind, reporting in the World Wildlife Fund's *Living Planet Report 2016,* writes that the population of vertebrates has declined 59 percent since 1970.[10] These losses are accelerating.

Although it is necessary to grieve, grief is seldom experienced in any straightforward way. Author, editor, and climate-justice essayist Mary Annaïse Heglar captures its nuance when she writes:

> I floated around on a dark, dark cloud. I frequently and randomly burst into tears, and I'd refuse to admit to myself that I knew exactly why I was crying. When I was around bustling crowds of people, I saw death and destruction. When I walked on dry land, I saw floods. I imagined wild animals, especially snakes, getting out of the zoos in the aftermath of natural disasters. I worried about how we would treat each other in the face of such calamity.[11]

I experienced something similar. When I was a child, my mother told me I could be "anything I wanted to be." I knew this wasn't literally true, but I also knew that I had many options. As I grew up, I set out to become a clinical psychologist, with plans to write books about psychology for popular audiences. I imagined myself married with children. What a lovely life I had planned! It was going to be meaningful, intellectually stimulating, financially rewarding, and rich in relationships.

But when I forced myself to face the climate crisis and to accept the truth of the climate emergency, I realized that my lovely life plan was a house of cards, ready to collapse at any moment. My life of privilege and self-cultivation would not be satisfying if it unfolded while tens of millions of refugees streamed out of regions made unlivable by heat, drought, or flood. Could I really focus on myself, my career, family, and interests, when state after state failed and collapse seemed imminent? Ultimately, I had to acknowledge that the future I had planned was ruined. I can't live a good, self-focused life while the world burns.

Do you feel that pain? I think that you do, though it may be too diffuse and unnamed to notice.

I specifically recommend you try to get comfortable with crying. This is often challenging, especially for those who have been taught that crying is a sign of weakness or that it signals an inability to cope. But crying is a specific act of emotional recognition and response; it is powerful, healthy, and necessary. It provides an outlet for all the grief and pain inside you, helping link the emotional and physiological. When you cry, you release toxins and stress hormones from your body.[12, 13]

Simply giving yourself permission to cry freely can be a tremendous relief and can allow you to gain access to other repressed or ignored feelings.[14, 15] Further, crying plays a critical social function, communicating to others, such as your friends and family, that you could use their comfort and support.[16]

We must decide that these losses deserve to be remembered, felt, and mourned. We must recognize that only by grieving these deaths and extinctions can we fully process our pain, honor our loss, and enable ourselves to engage in the reality of our already-diminished collective.

I must point out that grieving is not the same as giving up. There is a strain of climate "doomers" who say that humanity and the natural world are a lost cause. They believe that our Earth is in hospice and we should prepare it, and ourselves, for death.[17, 18] These doomers see their current calling as simply expressing their grief and persuading climate activists that they are high on "hopium."

While doomers understand the need for grief, they misunderstand grief's purpose. They use grief and loss as an endpoint, an excuse for inaction. But we can't *just* grieve. In this moment, we are called to do something very different: In the immortal words of labor organizer Mother Jones, "Pray for the dead and fight like hell for the living!"

Climate activists know that our mourning and loss is neither a reason to give up nor evidence that we have surrendered. Instead, these feelings reinforce our ties to humanity and life. The grief we feel reminds us that nothing is more precious than life. The grief calls us to embrace the most difficult challenge: to fight to save as much life as possible and—hopefully—to restore some of what's gone.

When you acknowledge that we are facing a climate emergency and allow yourself to feel the all-encompassing grief for what we've already lost, you also begin the vital work of re-establishing your visceral and intimate connection to all life. Even though we've been taught to view the natural world as fundamentally separate from humanity—a "resource" for our use—and to ignore our fundamental connection to the other living beings of the planet, our relationship to humanity and the natural world is our spiritual birthright. Macy cites Zen Buddhism to argue for the necessity of the connection:

> The Vietnamese Zen master Thich Nhat Hanh was once asked what we needed to do to save our world. "What we most need to do," he replied, "is to hear within us the sounds of the Earth crying." The idea of the Earth crying within us or through us doesn't make sense if we view ourselves as separate individuals. Yet if we think of ourselves as deeply embedded in a larger web of life, as Gaia theory, Buddhism, and many other, especially indigenous, spiritual traditions suggest, then the idea of the world feeling through us seems entirely natural.[19]

This is a very different view of the self than what Macy calls the "extreme individualism" that "takes each of us as a separate bundle of self-interest, with motivations and emotions that only make sense within the confines of our own stories."[20] When we throw out the lessons of neoliberalism and instead allow ourselves to face the truth of the climate emergency and grieve its associated losses, we begin to hear the Earth's story, which is a story of our interconnectedness.

Climate Activist Vanessa Nakate, founder of the pan-African Rise Up Movement, which demands an end to new fossil fuel infrastructure, and calls for climate finance support and disaster relief, told me something similar. After describing her experience visiting famine-stricken regions and meeting a young boy who died soon after, Vanessa told me, "I think all climate justice activism is about love. Otherwise, you would just act for yourself. Climate activism is motivated by love and hope."

You, too, must grieve. You must grieve the people and biodiversity we've already lost and the lives we lose every day. You must grieve the future you've dreamed of. Only then will you be ready to move into action. In this way, you can turn grief into power rather than stay stuck in paralytic despair.

By grieving what we have lost, we free ourselves to take joy in life again; not all climate emotions are negative.

For instance, I cherish my relationships with people around the world who are living in climate truth. I am so inspired by and proud of the movement's resurgence. When I see disruptive action in the news, demanding the end of fossil fuel expansion, I feel a rush of adrenaline and excitement. *Yes! We are finally fighting back.*

I feel a sense of deep connection with people living in climate truth, especially with activists who take risks, put their bodies on the line, rearrange their lives, or otherwise go all-in for the mission. These are my allies, my comrades, my people. Even though we may never meet, we are in this fight together.

I feel a loving connection to plants and animals, as well. Hiking in nature is an awe-filled, restorative experience. Having dogs has made me keenly aware of how much we have in common with other mammals, and how much love can be shared between species. The love that I feel motivates me to fight harder.

Sometimes, especially when surrounded by allies and when our work is bearing fruit, I am bursting with hope for a new world that is coming. I've seen so much change even in the short time I've been working on this issue. I feel humanity can and will respond to this crisis. Once we are in emergency mode, we can accomplish shockingly ambitious feats. I believe we will transform into a species that treasures and nurtures all life. When I look at history, I see extraordinary social movements that have sparked dramatic political shifts in consciousness and morality. I know what American industry and society accomplished during WWII. Then and now, we can wake up from the trance of denial and go all-in to fight an existential threat. I feel so proud to be a part of it.

You are probably hurting after reading all this, even if you believe that positive feelings and impacts may follow. I know this is hard, so I want to thank you for having the courage to face the truth with your heart and mind. It matters a great deal to our collective chances. If it is any comfort, facing reality and grieving the pain of it are the hardest parts of your transformation and of this book. It should get a bit less painful and a little more uplifting from here. Onward!

Questions for Reflection and Discussion

- How are you feeling after reading this Step?
- What emotional experiences resonated most for you?
- Which emotions are the most challenging for you to feel? Do you criticize yourself for having them and try to shut them

down? (For example, *Oh, don't feel angry and envious, you are such a bitch!* or, *Boys don't cry.*)

- Have you felt viscerally afraid of the climate emergency? Of what are you most afraid?
- Have you experienced grief for the people, species, and ecosystems already lost? If so, describe those feelings.
- What are your cherished hopes and dreams for the future? Have you grieved the fact that they may not happen as planned?
- Do you harshly judge your thoughts and feelings?

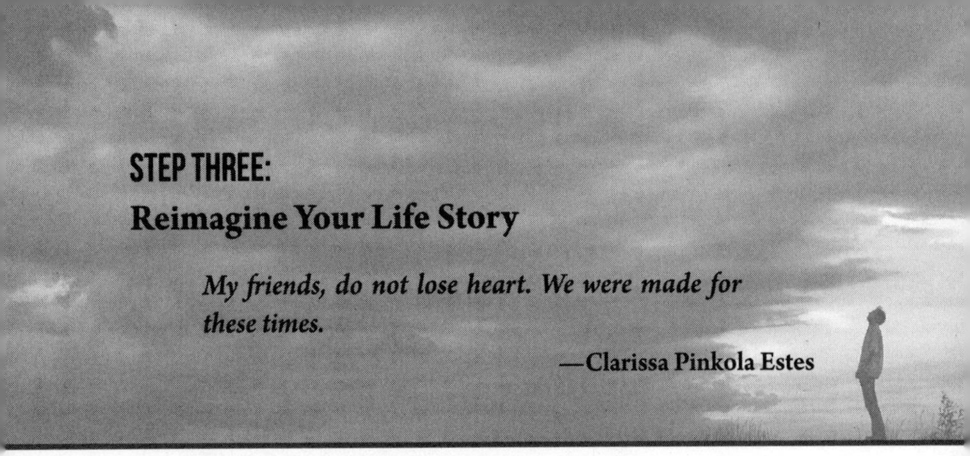

STEP THREE:
Reimagine Your Life Story

> *My friends, do not lose heart. We were made for these times.*
>
> —Clarissa Pinkola Estes

AT THIS POINT, you may be feeling the need to reconsider some fundamental things: Who are you? Why are you here? What is your purpose? The climate emergency challenges the answers we thought we had and invites us to revisit them. I discussed the need to grieve the future you thought you had. But what can replace it? In the context of the ecological emergency, we must each revise our story of self. In this new story, you are the hero. This designation might feel over the top. It might make you uncomfortable. But it's true: Humanity needs as many heroes as it can get—people who put the mission over their self-interest, people who realize that the mission is their self-interest.

What if everything in your life, including its most painful challenges, has prepared you for this role? You've experienced difficulties, but they've made you stronger. You've received gifts—maybe a nurturing home, a strong moral sense, a good education, financial security, or specific talents—but they were not randomly given. Everything you've experienced and gained has prepared you for your mission.

I know it's true for you because it's true for me, and it's true for so many others who have faced the truth and reoriented their lives around the climate emergency.

For instance, Zain Haq, a 21-year-old activist from Pakistan, is studying and protesting in Vancouver, Canada, and is at risk of deportation for his many arrests for nonviolent disruption:

> When I was 12, I witnessed an extreme amount of flooding in my country, Pakistan. I saw the displacement

> of millions of farmers and people fighting over bags of rations when they arrived in the city. From then onwards I knew that it is a binary choice between going "all-in" and being complicit in what is going on.... The best part of being an activist is the privilege and ability to live my truth and being able to speak and act in truth without fear. This is rewarding because it is contagious and inspired others to do the same in my experience.

What is this superpower that Zain has acquired, to "act in truth and without fear," while putting himself at risk? I call it "the movement mentality." It is a crystal clear understanding of your mission in life. It is not two hours every Thursday. It is an all-consuming commitment, born of desperation, which itself is borne out of our love for humanity and all life. I have experienced the movement mentality every day since February 2013, and I highly recommend it. I want to tell you a bit about the path that led me here, as it might be helpful in charting your own course.

I have been given many gifts that I strive to put into the service of the mission, including access to a world-class education, years of intensive psychotherapy, the socioeconomic privilege that has allowed me to work for years as a volunteer, and family and friends who believe in me and support me.

The challenges I have faced also taught me lessons that help guide me in this mission. My childhood prepared me to take responsibility for difficult problems, work through complex emotions, and shed preconceived ideas. My parents both came from troubled homes—my mother was the daughter of teen parents, and my father was the son of a depressed, traumatized Holocaust survivor. My parents valued the empathy and emotional insight I expressed as a youth and encouraged me to take on a leadership role in my family's emotional and social realms. For example, I remember them asking for input on how to deal with the challenges of raising my brothers, and taking my advice seriously. Perhaps this is why I have always been dubious of expecting authorities or anyone else to solve my problems. It has always felt natural to take on a leadership position, and I have always been encouraged to say, *Okay, I'll handle this.*

I've also drawn conviction from my family's background. I grew up hearing about the Holocaust from my grandmother. Whenever we spoke, whether on a visit, on the phone, or on a family trip, she talked almost exclusively about the Holocaust. She saw everything through its life-altering lens. She carried with her a deep and abiding feeling of betrayal—not just the betrayal she experienced by the Nazis, but the betrayal she experienced by ordinary Germans like her schoolteacher, who refused to acknowledge her on the street after my grandmother was kicked out of school. Hearing these stories at every visit instilled a visceral understanding of the adage that, *all it takes for evil to triumph is for good people to do nothing.*

I've also faced personal tragedy. My first love was my high school boyfriend, a brilliant actor, writer, and scholar who taught me how to play chess and, as my partner in our school's mock trial team, led us to win the state championship. We had been together for more than two years, thinking about our bright futures and considering attending college together, when disaster struck. He became floridly psychotic and experienced his first of many hospitalizations.

It is impossible to convey the overwhelming pain of watching a loved one go insane. In just a few days, my boyfriend went from being the most loved and trusted person in my world to being scary and bizarre, someone who didn't always recognize me. There was nothing I could do to help him. He was treated and medicated, but he wasn't the same and never would be again.

Neither would I. The pain I experienced watching him struggle in the grip of a disease he neither understood nor was able to effectively fight was a hurt I had never known before. College was supposed to be an exciting and emotionally connected experience.

I was supposed to be spreading my wings. Instead, I was alienated, traumatized, and depressed. During my first year, my boyfriend was in and out of the hospital; I felt guilt and shame for leaving him in such a terrible condition. I felt like the proverbial wolf caught in a trap, chewing off one of its legs to survive. I wondered if I would ever feel whole or good about myself again.

My boyfriend's life was immeasurably harder. His illness never relented, and each psychotic episode devastated him, to say nothing of the devastation his family and friends experienced. After struggling for over a decade with his disease, he reached a point of no return and killed himself. His illness and death left a gaping hole in the lives of everybody who knew him. He was a unique and beautiful person, overflowing with love and talent. He is irreplaceable, and his illness and death were unfair. I miss him terribly.

This experience taught me so many things, but the most important was simple: Terrible things—things you never thought to fear—actually happen. People may understand that bad things happen, but they usually think those things happen to other people. They may know that things go wrong, but they typically think that when things go wrong, they go wrong in expected ways. Only when someone has experienced a shattering tragedy or trauma do they realize that unforeseen, life-altering disasters can happen, leaving survivors with a deep feeling of vulnerability. I know in my bones that terrible things happen. My grandmother told me this over and over, but only by experiencing it myself did I internalize the message: Unexpected catastrophes happen, and we are all vulnerable to them. If we have a chance at preventing them, we must.

My experience with my high school boyfriend's tragedy taught me another lesson—what it feels like to possess a painful and uncomfortable truth. After his death, I received comfort from many people who reached out to me to tell me how sorry they were and that they were thinking of me. During his psychosis, however, I received little of this support—people didn't want to talk about what was happening to him. They felt uncomfortable. They didn't know what to say, and wanted to hope everything would be okay so that they could move on. I felt abandoned and alone.

Isolation is a common experience for people whose romantic partner experiences a psychotic episode. More than five years after I watched my boyfriend turn into someone entirely different, I was still searching for answers about what had happened, to him and to me. To understand my trauma, I developed my doctoral dissertation on the traumatic impact of psychotic episodes on romantic partners by interviewing women who had witnessed their partners' psychosis. Two key themes emerged from my work: the tendency of friends and family members to downplay the severity of an illness, and the estrangement and alienation the women felt because, in their words, "no one understood," and there was "no one to talk to."[1]

I have spoken with hundreds of people about their climate emotions over the years. The most common climate emotion that people express to me is loneliness, because "no one understands" or they "don't want to upset other people." This alienation makes it nearly impossible to process our emotions. They go unshared, unvoiced.

As we've seen with the denial surrounding our climate crisis, when an experience or phenomenon is taboo, shrouded in silence or shame, it becomes even more destructive. This is why, in Alcoholics Anonymous, they say that "you are only as sick as your secrets." In my interviews, many of the women were relieved and unburdened to be able to talk with someone about their experiences.

Brené Brown, an author and professor at the University of Houston, writes about the horrible consequences of shame. She encourages readers to have the courage to be vulnerable and share their lives with trusted others, especially the parts of life that make them feel ashamed. Our shame dissipates when met with empathy and understanding. Connection and trust take its place.

When I was a therapist, I worked with my clients to discuss the most painful parts of their lives—things that had made them feel ashamed—and I saw the healing power of taking painful secrets out of the shadows. Helping someone face and work through their personal crises or tragedy isn't that different from helping someone face and process the climate emergency. In both contexts, the helper must express empathy, curiosity, and a willingness to tolerate painful feelings in service of the truth.

Although my story is unique, we all have pain, challenges, and weaknesses, as well as blessings, privileges, and talents. As Longfellow wrote, "into each life, some rain must fall."[2] Take a moment to reflect on your own life: How have your gifts and challenges brought you to this moment, this book, and this cause? The reflection may reveal that you are more ready for this mission than you imagine. Maybe you have been preparing for it all along.

It's easy to see how the emotional and practical skills you've developed, the talents you've cultivated, the money you've saved, and the relationships you've built will assist you on your mission. But your challenges and most painful moments have also prepared you for this work. Perhaps a childhood illness taught you about the vulnerability of all things and your capacity for healing. Maybe being bullied led you to be assertive when necessary. Perhaps you experienced racism that taught you that the world is in desperate need of transformative change. Maybe the premature death of a loved one made you fiercer in your desire to protect others.

It has been redemptive for me to think of my experience witnessing my high school boyfriend's psychosis as preparation for the mission of protecting humanity and all life. For years, I felt bitterness and resentment. I felt that the world had been terribly unfair to me. I wished I could have a "normal and happy" life. But I don't want that anymore.

Because I experienced a catastrophic collapse up close, I can see a coming collapse that others are blind to. My experience made me more comfortable and motivated to break the silence and speak uncomfortable truths. While I am forever sorry that I couldn't protect or heal my boyfriend, I am glad to have an opportunity to prevent another calamity and help heal our society.

I am also grateful that my grandmother taught me about catastrophe at a young age, including giving me a moral framework around it. She taught me that it is not morally acceptable to be passive in the face of imminent disaster. We cannot be the "good Germans" who supported the Third Reich's atrocities in their passivity and inaction. We cannot make "fitting in" or "not rocking the boat" our priorities. Instead, we must uphold our values—social norms be damned—and fight for humanity and the living world.

There is something else on your side, too. While I considered myself an atheist for most of my life, my climate work has given me a profound spiritual sense. I believe there is an unseen, powerful force on our planet that wants to live. It's a force that exists inside all of us: It's why our hearts beat without our volition. It's the reason we're able to breathe without conscious effort. This vital force ensures that we, along with plants, other animals, and even microbes, seek out hospitable environments in which to grow and reproduce. The force that wants to live could be a product of evolution—plants and animals that don't display a drive to eat, drink, and avoid predators die.

This life force also manifests in our *collective* desire not to protect just our own lives but to protect *all* life. Different cultures have recognized this essential component of human experience and expressed it in different ways—"Tao" in Chinese, "Tathāgatagarbha" (or Buddha-nature) in Sanskrit, "Ruh al-Qudus" in Arabic, "Chai" in Hebrew, "Holy Spirit" to Christians, "The Power of Love" to Martin Luther King Jr., and "Spirit" to the Water Protectors at Standing Rock.

Another name for this force is *negentropy*. In his 1944 book *What Is Life?* Erwin Schrödinger notes that everything in the universe is subject to the second law of thermodynamics, *except life.* Jeremy Lent (2022) explains:

> This loophole was first described in 1944 by Austrian physicist Erwin Schrödinger, who won a Nobel Prize for his work on quantum theory, then ventured into the fundamentals of biology. In his seminal book *What Is Life?* Schrödinger explained that living organisms exist by converting the entropy around them into order, creating temporary eddies of negative entropy, which he called negentropy. They're not exactly repealing the Second Law because, as they organize the energy and matter within themselves, they're increasing entropy in the universe as a whole. But it's a local loophole to the law that's maintained itself on Earth for billions of years. Wherever there is life, entropy is being reversed—at least for a while.[3]

Life turns disorder into order. We do so through our very metabolism—living things take in nutrients and transform them into a living being. Space is vast and dark and cold, but Earth is still teeming with life, because of this unique ability.

In his book, *The Social Conquest of Earth,* biologist E.O. Wilson describes how evolution shaped our desires, writing that humans evolved both to maximize their individual self-interests and to protect the group and ensure it thrives. These two forces—individual selection (our desires to protect and enhance ourselves) and group selection (our desire to protect and enhance the group)—have been vital in shaping human evolution:

> The dilemma of good and evil was created by multilevel selection, in which individual selection and group selection act together on the same individual but largely in opposition to each other. Individual

> selection is the result of competition for survival and reproduction among members of the same group. It shapes instincts in each member that are fundamentally selfish with reference to other members. In contrast, group selection consists of competition between societies through direct conflict and differential competence in exploiting the environment. Group selection shapes instincts that tend to make individuals altruistic toward one another (but not toward members of other groups). Individual selection is responsible for much of what we call sin, while group selection is responsible for the greater part of virtue. Together they have created the conflict between the poorer and the better angels of our nature.[4]

Wilson points out that altruism forged by group selection is only extended to the "in" group. But that won't work this time. This time, we need to all act together for the good of all. We need to recognize that we are *all* in the "in" group. I believe that the force on Earth that wants to live is pushing us beyond that boundary toward this realization of interconnectedness and the individual belief that each person's safety and well-being is intertwined with the biosphere's safety and well-being. As Martin Luther King Jr. wrote in "Letter from a Birmingham Jail": "We are caught in an inescapable network of mutuality, tied in a single garment of destiny."[5] We will only be able to truly thrive on a healthy planet when enough of us realize that we must take our responsibility seriously to protect and nurture the natural world, and, by doing so, protect and nurture each other.

The vital force is the liveliness we feel in ourselves and our sense of connection to and love for other living things. It is the part of us that feels responsible for other people and species. It is the part of us that makes existence worthwhile for its own sake.

It is the part of us that makes us more than mere self-interested machines programmed to consume, compete, and build wealth and power. It's the part that makes us unique creatures capable of protecting and contributing to the greater community of life.

As Wilson points out, there are other forces within us, too. The drives for greed, separation, and narcissistic gratification are also inside us. Buddhism regards greed and hatred as mental afflictions that veil and obscure our true life-affirming nature. Our consumerist culture and alienating political economy have nurtured these destructive forces and quieted the force on Earth that wants to live, regarding it as naïve and undisciplined.

The consumptive, narcissistic, destructive drives have been the dominant ones in American culture for decades. But—and here's the good news—the climate movement's growth suggests that the better angels of our nature presiding over the interconnected, life-loving, protector forces are resurgent. I believe that the force that wants to live is making a last stand as the climate crisis approaches the point of no return, likely because it faces the same extinction as the rest of us: If we die, it dies. But it is not going softly into the good night. Instead, it is rising up in millions of us, commanding us to drop everything and fight for life.

Once you listen to this force and decide to go all-in for all life, the force becomes an active—if silent—partner. Since I decided to do everything I could to protect humanity and the natural world, the wind has been at my back. I have been helped by hidden hands: The right people have appeared at the right time and helped push me forward.

I didn't suddenly feel the life force all at once, however. When I started to read more about the climate emergency, the climate dominated my thoughts. It felt as though sirens were blaring inside of me, a bright red flashing light in my face. I had nightmares about huge waves. I felt raw terror paired with grief and sorrow. I had trouble focusing on the issues that had previously occupied my mind, such as my academic and career success, my patients, my social life, my appearance, or my apartment. It came to a point where I could think only about one thing: the impending disaster.

I needed to do something. I started to write and publish, but it didn't feel like enough. It wasn't until I felt the life force in me that I began to hear more substantive answers. I clearly remember the moment I began to hear them most explicitly. On a winter evening in 2013, my friend Ryan Peterson asked me to set my sights higher than I would have ever set them myself: "Discourse isn't enough," he said. "Think—what could you do or what could we do together that could actually *solve* the climate crisis?" His words resonated in me, echoing at a visceral level. It finally clicked: Not only am I connected to all of life, but that I can—and have to—take responsibility for protecting that miraculous web of connection.

I had never let myself think so big; I had imagined I could offer incisive commentary, not that I could try to *solve* the crisis. Yet, once Ryan issued the challenge, I never wanted anything more. The force on Earth that wants to live was coursing through my veins.

Whereas the climate emergency once filled me with terror and despair, the challenge to "solve" the emergency lit a fire in me that has never dimmed. I have established a long-term partnership with the force on Earth that wants to live; I have joined its team and become one of its agents. In the process, I have changed drastically. My identity, priorities, my spiritual beliefs, what makes me feel good about myself, and how I spend my time. For example, I always found fundraising anxiety-producing and awkward. And yet, I now spend most of my time doing it, as joyfully as possible. Why? Because I want to do whatever I can to help the movement. That's really all that matters. I have the movement mentality.

By fighting for humanity and all life, we channel the forces within us of *both* individual and group selection. We become agents of negentropy. We channel the force on Earth that wants to live. We realize that, because winning total mobilization of our economy and society to restore a safe climate while we feed and house everyone on the planet is our only hope, supporting the movement is in our self-interest. We are transformed by the responsibility of our mission.

Questions for Reflection and Discussion

- How has your life prepared you to face the climate emergency?
- What challenges have you faced that have prepared you to: tell the truth, be resilient in a catastrophe, or lead or help others?
- What have you learned from your parents, grandparents, and other ancestors that has prepared you for the climate emergency?
- Have you ever experienced or witnessed a catastrophic collapse? Did someone you know suffer a psychological breakdown? Did a neighborhood burn down, an organization destroy itself, a society descend into chaos? What have you learned from those experiences?
- Can you envision a life that revolves around a commitment to protect all life?
- In what ways do you feel unprepared? What do you need to work on? What do you need help with?
- Does the concept of a "force on Earth that wants to live" resonate with you? Do you feel that force inside of you?
- Do you have a spiritual practice and perspective that equips you to protect humanity and all life?
- What are your greatest fears about becoming a climate hero? What are your hopes?

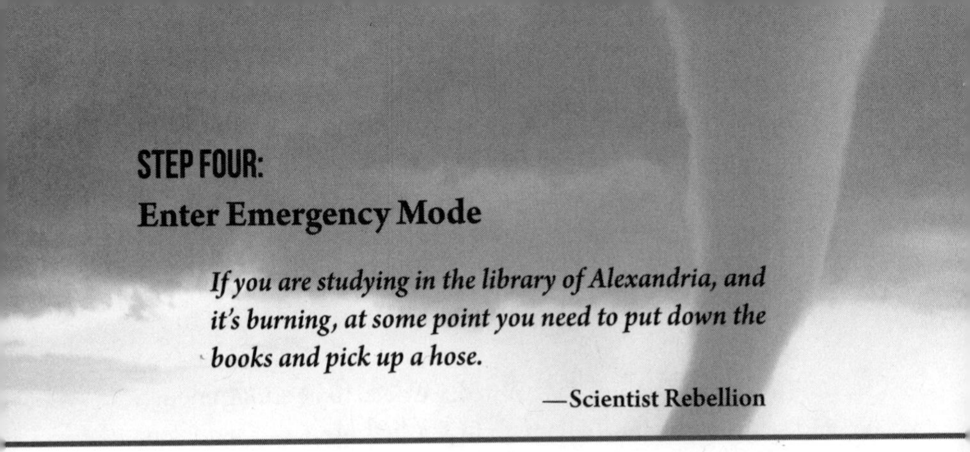

STEP FOUR:
Enter Emergency Mode

> ***If you are studying in the library of Alexandria, and it's burning, at some point you need to put down the books and pick up a hose.***
>
> **—Scientist Rebellion**

IF THE PRECEDING MATERIAL unsettles and terrifies you, good. We are in terrible danger—you and me, my family, your family, and the whole human family. So let's act like it, and get busy protecting everyone we love. Feel the fear, grieve the future, and let it motivate you to action.

We can rapidly transform our economy and society to beat back a global catastrophe—I know because we've done it before. But to solve an emergency like the climate crisis, we must collectively and immediately exit "normal mode" and abandon the gradual policy advocacies and enervated emotional states that accompany it. We need a collective awakening on the scale of our response to a national attack. Together, we need to realize, in accordance with Wilson's theory of group selection, that we face an existential emergency, that we are in clear and imminent danger, and that we must immediately mobilize with everything we've got.

We can only do this when we enter emergency mode. In emergency mode, we can channel our fear to fight the threat facing us collectively and creatively. This is the opposite of panic mode, in which we freeze or take flight. When we enter emergency mode, we respond to threats with thoughtfulness, planning, and coordination. Our senses are heightened; we feel a preternatural calm; we are able to process large amounts of information quickly to arrive at a flexible course of action.

When a society enters emergency mode, it mobilizes and works collectively to address and solve huge problems quickly.

We know we can mount this response because we've done something similar before. By looking to history, we can see 1) that our country and our world need emergency mobilization, and 2) that an assertive mass movement that builds power is the way to get there.

To take this path, as I've argued, we need to be strong enough and self-compassionate enough to *feel* the threat and to *feel* our fear. As individuals and as a society, we need to use this fear to wake us up. You already know that it starts with you, looking inward—feeling the unbearable urgency of the crisis in your bones, taking responsibility for solving the climate emergency, committing to fighting for all life, and stepping up to turn that commitment into high-leverage action.

In 2016, I published a paper arguing that the climate movement needed to show, through its language and tactics, that it had exited normal mode and was operating with a fundamentally different level of urgency. Thankfully, due to the actions and advocacy of many, this approach has been adopted by significant elements of the climate movement, such as Extinction Rebellion, Scientist Rebellion, Just Stop Oil, and the School Strikers—exactly the kinds of groups Climate Emergency Fund now supports.

Emergency mode describes how individuals and groups function optimally during an existential crisis. When individuals and groups enter emergency mode, they position themselves to achieve incredible feats through intensely focused motivation and collaboration. I described the movement mentality that I and other activists experience: That's one form of emergency mode.

Emergency mode is a fundamental departure from a so-called normal mode of functioning. In normal mode, the individual or group feels relatively safe and secure. Immediate existential or major moral threats aren't recognized, either because there are none or because the vast majority of people are in denial.

	Normal Mode	Emergency Mode
Priorities	Many balanced priorities	Overriding priority: Solve the crisis
Resources	Distributed across priorities and saved for the future	Huge allocation of resources toward the solution
Focus	Distributed across priorities	Laser-like focus
Self-Esteem Source	Individual accomplishment	Contribution to the solution

Climate Emergency Unit—another organization Climate Emergency Fund has supported—lays out the four characteristics indicating that an institution, such as a government or a business, is entering emergency mode:

- It tells the truth about the scale of the crisis.
- It sets up new institutions to handle it.
- Voluntary or incentive-based measures become mandatory measures.
- It spends what it takes to get the job done.

This is a great checklist for evaluating governments, companies, and NGOs that claim to be treating climate like an emergency. The vast majority are not even close. There are grassroots climate organizations popping up all the time, all over the world, that operate in emergency mode, but precious few legacy institutions have been able to reorient toward it.

However, we are seeing visionary leadership in some quarters. For example, rather than sell Patagonia, worth $3 billion, Yvon Chouinard wanted to ensure that *all* Patagonia's profits, approximately $100 million per year, go to fighting climate change through funding grassroots movements and political candidates.[1] In order to facilitate this mandatory use of profits for climate, Chouinard set up a new institution, Holdfast Collective, a 501(c)(4). An estate planner for the ultra-rich commented that he had never seen anything like it.

The outdoor clothing retailer is an example of what it is possible to achieve when someone enters emergency mode or has the movement mentality. Its rarity also illustrates how drastically far away most businesses and institutions are from this goal.

One way to think about emergency mobilization is that it's both cause and consequence of a whole society experiencing the movement mentality. And it's a beautiful thing. People ask not, "How do I improve my own life?" but rather, "How can I contribute to the cause?" The mentality can foster an increase in societal trust and a stronger sense of national purpose.[2] For those who accept the need to rapidly—not gradually—convert an economy to a new purpose, emergency mobilization is the most effective, egalitarian, and sensible approach.

Historian David Kaiser describes the public's emergency-mode mindset during WWII: "At no time in American history have they shown more willingness to make financial sacrifices to meet common necessities, largely because they agreed with their president that the survival of civilization was at stake."[3]

What are the mental health effects of living in a society where everyone is in emergency mode, together? During the COVID-19 pandemic, some parts of the world experienced this "pull-together effect." For example, suicides dropped slightly in the United States in 2020 compared to 2019, and far more dramatically in Canada, which had higher levels of social trust and less political division over pandemic response.[4, 5]

Historically, the experience of Londoners during the Blitz is instructive. Journalist Sebastian Junger looked at how the Blitz—the nine-month bombing of London that killed 43,000 people and terrorized 8 million—affected the Londoners who experienced it. The government expected widespread panic and psychological breakdown, given the widespread trauma. But instead, Junger writes:

> Psychiatrists watched in puzzlement as long-standing patients saw their symptoms subside during the period of intense air raids. Voluntary admissions to psychiatric wards noticeably declined, and even epileptics reported having fewer seizures. "Chronic neurotics of peacetime now drive ambulances," one doctor remarked. Another ventured to suggest that some people actually did better during wartime.... Researchers documented a similar phenomenon during civil wars in Spain, Algeria, Lebanon, and Northern Ireland.[6]

In *A Paradise Built in Hell,* Rebecca Solnit explores the Blitz and several other examples of how, during disasters, communities strengthen and people come to each other's aid. Paradoxically, it is often fulfilling to live through them. Solnit describes:

> Strangers become friends and collaborators, goods are shared freely, people improvise new roles for themselves. Imagine a society where money plays little or no role, where people rescue each other and then care for each other, where food is given away, where life is mostly out of doors in public, where the old divides between people seem to have fallen away, and the fate that faces them, no matter how grim, is far less so for being shared, where much once considered impossible, both good and bad, is now possible or present, and where the moment is so pressing that old complaints and worries fall away, where people feel important, purposeful, at the center of the world.[7]

The US and UK home front mobilizations are powerful examples of what can happen when a society enters emergency mode. It differs markedly from business as usual or normal, aka political paralysis mode, as David Spratt and Philip Sutton outline in their groundbreaking book, *Climate Code Red: The Case for Emergency Action*:[8]

> Emergency mobilization is an economic approach that directs the collective force of industry away from consumerism and toward a singular national purpose. Profit-seeking activities are channeled toward the national mission and are tightly controlled if excessive. Emergency mobilization is characterized by large-scale deficit spending; sweeping command-and-control regulations, a high degree of citizen participation; increased taxation, especially of the wealthy, to control inflation and raise revenues for the national project; and robust government control over the allocation of raw materials and essential goods. And although corporations can play a constructive role in implementing mobilization, they do not drive the change process.

This type of all-in, maximum-intensity emergency mobilization is the only approach comprehensive enough to resolve a crisis on the scale of the climate and ecological emergency.

Ezra Silk, Climate Mobilization's cofounder and director of strategy and policy, wrote *Victory Plan* to demonstrate what a 10-year national emergency mobilization plan to address the climate and ecological crisis could actually look like. *Victory Plan* outlines an emergency mobilization that aims to:

- Create a zero-emissions greenhouse gas economy in ten years and draw down excess greenhouse gases until a safe climate is restored.

- Shrink the ecological footprint of the global economy from 1.7 planets per year to approximately half an Earth per year.
- Halt the sixth mass extinction by creating an interconnected global wildlife corridor system covering half the Earth's land surface and oceans.
- Create a society and global economy that works for everyone, including a healthy and stable international environment, healthy food, clean air and drinking water, life-affirming work at a living wage, medical care, housing, and full democratic participation in government and the workplace.

Victory Plan includes policy recommendations, such as an immediate ban on all new fossil fuel infrastructure and a 10-year, government-coordinated phase-out of the oil, coal, and gas industries through a rapidly declining cap on fossil fuel extraction and imports. It details massive energy conservation projects, renewable energy investments, and a new super-smart electricity grid. It calls for banning factory farms, transitioning off pesticides, and launching a national program supporting regenerative agriculture and plant-based diets. *Victory Plan* bans single-use plastic, fostering a circular economy where nothing is wasted. It includes reforestation and rewilding projects to restore biodiversity. All hands will be on deck—the federal government will guarantee a job to anyone who wants to take part in the mobilization. Others can participate by cutting their home energy use, biking, growing food in Victory gardens, and participating in community efforts for preparedness. Everyone will have a role to play.

The vision articulated in *Victory Plan* has already inspired others to raise their expectations of what's possible. It has influenced the policy vision proposed by Representative Alexandria Ocasio-Cortez and the Justice Democrats in their Green New Deal legislation, which calls for a 10-year WWII-scale mobilization and describes a range of programs to ensure racial, gender, social, economic, and environmental justice. These measures would also provide tremendous stability for displaced workers and all Americans during rapidly changing times.

Ezra and I had the honor of working with Congresswoman Ocasio-Cortez, Senator Bernie Sanders, and Congressman Earl Blumenauer on a national Declaration of Climate Emergency into Congress, introduced in 2019, which would put the House and Senate officially on record acknowledging the existence of a climate emergency and calling for a massive-scale national mobilization that phases out oil, coal, and gas and reverses climate change at emergency speed.

Victory Plan calls for emergency mobilization because there is no faster, more powerful way to transform our economy. Our grandparents mobilized to beat back an existential threat, and we can, too. Once we have fully entered emergency mode and mobilized, we will be optimally positioned to shatter every record and expectation in order to restore a safe climate and threatened ecosystems, and create a society based on meeting human needs and protecting all life.

But *how* we make this happen is the defining question of our moment. It's the question that should keep everyone who is living in climate truth up at night. The need for WWII-scale climate mobilization is shared by many notable intellectual figures, including economists Jeffrey Sachs and Joseph Stiglitz,[9] *New York Times* columnist Thomas Friedman,[10] author/ activists Naomi Klein[11] and Bill McKibben,[12] and policy expert Rhiana Gunn-Wright.[13] But having these luminaries onside isn't enough. *Knowing* or even *arguing for* what is necessary doesn't mean it will happen.

To solve an emergency like a world war or climate crisis, we must collectively exit normal mode. We each need to enter emergency mode personally and then communicate our choice to others. As I've already argued, how we respond to threats—whether by practicing denial and remaining in normal mode or by facing our fears and entering emergency mode—is *highly contagious*. We can initiate collective behavior by transforming as individuals and leading others.

According to psychologist Daniel Gilbert, humans are wired for a reflexive response to threats that are "intentional, immoral, imminent, and instantaneous."[14] The attack on Pearl Harbor was marked by all these characteristics. The shock and feeling of betrayal created a recognizable and collective sense of fear and vulnerability, as well as anger and a fierce desire to *respond* by fighting back. Plus, the groundwork had been laid by the preparedness campaign led by President Roosevelt, a group of forward-thinking government and military officials, and business leaders who had already started the rearmament processes.

Some assume that increasingly apocalyptic fires or storms will create a "Pearl Harbor moment" for the climate emergency. But climate change is not perceived by the public as intentional, immoral, imminent, or instantaneous—so we cannot expect the public to switch into emergency mode on their own. We have to lead them there.

Social movements have successfully illustrated the efficacy of emergency mode, too. AIDS Coalition to Unleash Power (ACT UP), an activist organization working to end the AIDS pandemic, shows how a citizen group can actively work to change a government response and provoke sweeping change. Through its actions, ACT UP made the federal government treat the AIDS epidemic like the public health emergency it is.

In the 1980s, HIV, the virus that causes AIDS, was spreading at horrifying speed and decimating gay communities in New York, San Francisco, and other large cities. The government, as a result of pervasive homophobia, gave HIV patients no help, and neglected research and treatment. The government's failure to act destroyed whole communities.

It took now-iconic AIDS activist Larry Kramer to channel the terror and anger of the gay community and their allies into a social movement capable of demanding an appropriate response from the government and other institutions. Although Kramer co-founded the Gay Men's Health Crisis (GMHC), he broke with the group over disagreements about strategy and tactics in responding to the AIDS crisis.

According to Kramer, the GMHC did not enter emergency mode. Instead, it continued to seek solutions through insider channels, holding meetings with government officials to ask for help. These strategies were not working. Kramer criticized GMHC, arguing that it was so focused on claiming mainstream status that it was helping people die rather than fighting to protect the living. In response, Kramer co-founded ACT UP, a nonviolent but militant organization that viewed AIDS as an existential threat. ACT UP did not politely ask the government for help: It *demanded* emergency action.[15]

Kramer knew he was fighting for his own life and the lives of his friends. He had no interest in business as usual. He wanted the government to act on AIDS *now*—to research the illness, find treatments, treat the sick, and prevent transmission. Kramer treated AIDS with deadly seriousness, using fear and inflammatory rhetoric to provoke people to action. He urged his would-be supporters to feel as much fear as possible—telling crowds of gay men that if they didn't fight back, they would soon be dead. Kramer referred to AIDS repeatedly as a "plague" and to the politicians who ignored it as "Nazis" and "murderers."[16]

The symbol of ACT UP, a pink triangle, references the genocide of gay men during the Holocaust. Its slogan, "Silence = Death," referred not only to the government and media's silence on AIDS but to the cultural silence around homosexuality. At the time, many gay people were closeted, hoping to avoid discrimination

CREDIT: ACT UP

and dehumanization from a homophobic culture. Kramer's solicitation of fear and bracing language and symbology was tactical: He was not just inviting other gay men to join him in emergency mode and to focus intensely on solving the crisis, he was arguing that there was no alternative.

The silence around homosexuality, with most people keeping their sexual orientation at least partially private, posed a huge problem for the movement. It contributed to the ignorance about AIDS. Gay people, including gay government workers, gay researchers, gay doctors, and others, could not work together with maximum impact or collectively communicate the emergency to the public while still in the closet. The public assumed that if they weren't hearing much about a health crisis, there must not be one. Kramer spoke to this in his landmark essay, "1,112 and Counting."

> Why isn't every gay man in this city so scared shitless that he is screaming for action? Does every gay man in New York want to die?... I am sick of closeted gay doctors who won't come out to help us....I am sick of closeted gays. It's 1983 already, guys, when are you going to come out? By 1984, you could be

> dead. Every gay man who is unable to come forward now and fight to save his own life is truly helping to kill the rest of us. There is only one thing that's going to save some of us, and this is numbers and pressure and our being perceived as united and a threat. As more and more of my friends die, I have less and less sympathy for men who are afraid their mommies will find out or afraid their bosses will find out, or afraid their fellow doctors or professional associates will find out. Unless we can generate, visibly, numbers, and masses, we are going to die.[17]

ACT UP also held disruptive protests to demand that the government launch a crisis response to the AIDS epidemic. They channeled their grief and terror into effective activism; they used it to build a movement. In *This Is an Uprising,* author-activists Mark and Paul Engler describe ACT UP's hard-hitting and creative approaches: activists chaining themselves to buildings; blocking FDA offices and papering the offices with posters of bloody handprints; stopping traffic in high-profile places like the Golden Gate Bridge; interrupting CBS's nightly newscast with on-air protests; and spreading ashes of dead loved ones on the White House lawn.[18]

ACT UP mobilized an incredibly effective response to the HIV/AIDS crisis in the United States, forcing the government to mobilize its public health and research apparatus, winning funding for research and the right of AIDS patients to participate in every phase of the drug development process. ACT UP's success also laid the groundwork for mainstream acceptance of homosexuality in America and gave energy to the continuing struggles for gay rights and equality everywhere.

ACT UP's success was incomplete. AIDS became a global pandemic, and one million people still die from AIDS around the world every year. But researchers, doctors, nurses, policymakers, public health officials, journalists, government officials, and more worked tirelessly for more than thirty years to create today's conditions, where more than 23 million people globally are receiving highly effective antiretroviral treatment,[19, 20] there is preventative PREP medication available for those at risk, and people diagnosed with HIV can live long, healthy lives. ACT UP instigated this by leading individuals and institutions like local, state, and national governments, the National Institutes of Health, hospitals, and universities to treat HIV/AIDS like the crisis it is.

There is so much to learn from ACT UP—their tactics, of course, but also their mentality, and their focus on breaking the silence. ACT UP is one of the best examples we have of the movement mentality, combining the recognition of an emergency with the understanding that the only solution is collective. Self-interest is merged into the success of the movement. We realize that nothing is more important than the cause and do truly remarkable feats in its service.

Movement historian Sarah Schulman describes:

> They (ACT UP activists) came to save lives with humor, commitment, profound innovation, genius, will, and focus, and sometimes wild acting out, ruthlessness, and chance. But that meant that they also came to die and to watch disintegration. Because, to make something better, we have to face it at its worst. And only a small group of people on this Earth are willing to look pain in its real face, assess it accurately, listen, and then criticize themselves with rigor, find a productive way to cooperate, and rise to the occasion to solve a problem. People who are desperate are much more effective than people who have time to waste.[21]

Schulman and I are in resounding agreement regarding the need to face the truth and assess the strategic situation. But the other elements of her description are fascinating, as well. To succeed, activists need to 1) listen, 2) criticize themselves with rigor, and 3) find a productive way to cooperate. Accomplishing this is what allows us to rise to the occasion to solve a problem.

Anyone who has worked in an office, had a family, or attended a city council meeting knows that listening and working productively together are hard things to do. And in movements, the people you strive to work productively with are usually volunteers and often strangers. It should be no surprise that activists often identify interpersonal conflict as one of the biggest challenges of participating in a movement.

The element of rigorously criticizing ourselves is also critical. This is not about meanness, masochism, or purity politics but rather about continual improvement in efficacy and strategy. The ability to self-assess, as an individual and an organization, contains tremendous power. It is not enough to "do something" in the face of the climate emergency. We have to rigorously and continuously ask ourselves, What can I do better? How can I have the greatest impact? And, What can our group do better? How can we have the greatest impact? These are touchy questions. No one likes to have their action, plan, or idea criticized—I know I don't. But if we are in the movement mentality, we put the success of the movement above our ego. This is an emergency! This is a race against time! We *have* to ask tough questions.

It is irresponsible to say, for example, "Just do *something* on climate," and then praise any action or campaign under that banner. That's the nice and polite approach, and it won't anger anyone. But it's not what we need. We have a moral and strategic obligation to rigorously and relentlessly improve—to grow our power, efficacy, and impact. While the climate emergency is accelerating, we simply can't settle for "pretty good" because, as Bill McKibben says, "Winning slowly is the same as losing."[22]

It's a miracle that any movement succeeds, given all the challenges most movements are up against. This is why desperation is such a key driver: Activists have to be so committed to the cause that they overcome the significant hurdles thrown in their way. When they enter emergency mode, it's because they have realized that the *only* way to safety is through a collective solution. This is the movement mentality. Activists with the movement mentality are obsessed with the mission and willing to overcome any internal or external obstacle to win.

ACT UP also pushed coming out of the closet without shame. This was an incredibly important and successful strategy. Silence is devastating politically because if you are hiding, you cannot raise your voice. How many people came out in response to the AIDS crisis? How many individual conversations were had among families, friends, and colleagues? Perhaps tens of millions. Learning that people they loved and respected were gay and in danger profoundly impacted many people. Homosexuality and AIDS could no longer be abstract phenomenon that affected "other" people. Breaking the silence brought the crisis home.[23]

Like ACT UP, the immigrant rights movement also struggled with potential activists not joining the movement or actions because of their fear of consequences. In order to succeed, the DREAMers couldn't stay in the shadows. So they broke their silence, "coming out" publicly as undocumented at protests. This put them at risk, but some risks are worth taking. As Jonathan Perez, a 23-year-old immigrant rights activist put it, "We need to live without fear because the fear paralyzes us. If we stay quiet, we stay in the shadows."[24]

It's time for each of us to break the silence about the climate emergency—to tell the truth, loudly, and all the time. *Talk about climate* is the one mode of engagement that I recommend to everyone. It's healthy, for you and your relationships, and it is highly politically effective. Start with your friends and family and expand from there. A recent study from Yale University confirmed the impact of this approach, finding that "discussing global warming with friends and family leads people to learn influential facts, such as the scientific consensus that human-caused global warming is happening. In turn, stronger perceptions of scientific agreement increase beliefs that climate change is happening and human-caused, and increase concern about climate change."[25]

If we are silent, we are powerless. Even today, despite the moment of whirlwind we inhabit, most people remain silent. The Yale program on Climate Change Communication[26] found that only 9 percent of Americans hear people they know talk about climate change at least once a week, and only 15 percent once a month.[27]

Yet the same study found that 35 percent of Americans are "very worried" about the climate, and another 35 percent say they are "somewhat worried." This is pluralistic ignorance in action. Seventy percent of Americans are worried, but they aren't talking about it. Instead of collective political will, they perpetuate the illusion that "everything is fine," while feeling alone with their dismal knowledge.

During lockdown in 2020, I created Climate Emotions Conversations with the help of Canadian climate activist Jessie-Ann Baines. Climate Emotions Conversations is a free virtual platform where you participate in a guided video call with a small group of others. You each take turns sharing and listening to each other's climate feelings. Hundreds of people have joined these conversations, from the US and all over the world.

The Climate Emotions Conversations are incredibly powerful. Through sharing your feelings and hearing from others, you see that climate grief and fear are not rare at all. They are not alien, but deeply human. As one participant described, "It was beautiful and nourishing and honestly a little bit world shifting. I am buzzing with energy of feeling seen and empowered and connecting with others who lead lives so different from my own.... My mind is blown, humans really are something incredible."

Harnessing the feelings generated by Climate Emotions Conversations is our best hope for transformation. That's why signing up for a free sharing and listening session is the ideal next step to take once you've finished this book. Other supportive organizations available to help you talk about the climate emergency and dive even deeper into exploring your own experience include Good Grief Network,[28] which runs a 10-step program for small groups to discuss and process the climate emergency together. You can also host or attend a Climate Cafe.[29] Or check out the community around Joanna Macy's work, the Work That Reconnects Network.

While it may be easier to discuss your feelings with strangers, it's past time to let your friends and family know how you feel about the climate emergency. Perhaps you worry about the social awkwardness of bringing up the climate emergency—and it may be awkward at first. But keep in mind that many of your friends and family are also worried, and they will be relieved and appreciative when you bring up your feelings, especially if you can offer them support and guidance. Be personal, be emotional, be authentic and empathetic. Hear people out and make them feel listened to. Share this book, or any other climate book that has moved you, with others. "I know we've shared our concerns about the climate crisis, and this book has really helped me. Would you be willing to take a look and tell me what you think?" You can also invite them to a Climate Emotions Conversation.

Once you're comfortable talking about the climate emergency with your friends and family, begin to talk with others. Set a goal to talk about the climate emergency and the need for climate mobilization once a day; then, build up to talking about it more than once a day. Consider showcasing the climate emergency message on T-shirts, hats, pins, or bags. This is another way you can communicate your affiliation with the movement and invite potential conversations with others.

Have these conversations online, too. You can talk about the climate emergency and the need for mobilization on social media and, depending on your access, on email lists, blogs, or in mainstream publications. How many followers do you have on TikTok, Instagram, or Facebook? Do you use Reddit or Medium, or make YouTube videos? You may be able to make a significant impact by spreading climate truth through those channels.

Social movements have long utilized cutting-edge communications technologies that have not yet been controlled and co-opted by the powerful to fight denial and spread their messages. In his time, Martin Luther used the revolutionary potential of a new technology—the printing press—that led to the Protestant Reformation. Back in 1518, printing hundreds of copies of your political arguments and distributing them was an innovation, and a very effective one.[30]

Half a millennium later, the Civil Rights Movement used television as a tool to broadcast the violence of segregation into millions of American homes.[31] Civil disobedience created hundreds of dramatic, suspenseful scenes, such as confrontations during lunch-counter sit-ins. The public was captivated. What would happen? How would this turn out? How would the owner and waitstaff respond to this protest? How would the protesters respond to abuse? Would law enforcement get involved? Would there be violence? Would people die? For many, these scenes unfolding on the news night after night was a spear in denial's heart. Think the system isn't so bad? Look what happens to those who challenge it. The brutality and oppression inherent in the Jim Crow system, as well as the dignity and humanity of African Americans, were brought by television into Americans' living rooms.[32]

For climate activists, the Internet is our cutting-edge technology. Social media offers the possibility for social movements to burst into the forefront of the public consciousness.

It happened with #MeToo: An oppressive silence was broken, and powerful men finally began to be held accountable for sexual harassment. The social media-based #MeToo movement illustrates, once again, the centrality of integrating the intellectual and emotional, the personal and political. Women sharing en masse their stories of sexual harassment made many men finally realize the huge scope of this problem, and made institutions realize that they needed to act.

Black Lives Matter also demonstrated the power of social media to instigate street action and cultural political change. The Black Lives Matter movement started as a hashtag on Twitter. That is the potential catalyzing power of language on social media. Those three words captured the painful reality that needs to be faced—that our policing and carceral system are racist, unjust, and destructive—but phrased it in an affirmative way. #BlackLivesMatter is both an indictment, that it needs to be said at all, and a validation. This hashtag coalesced all that meaning into three words, social media took it viral, and *then* we saw massive street protests.[33] That is the immense power of social media—if the message is truly resonant.

Imagine if a #NoFuture or #LastGeneration movement started about the climate emergency, with first a handful, and then hundreds, thousands, and even millions of people sharing their personal reactions to the climate crisis—expressing their fears and revealing their dashed hopes for their futures. This would be synergistic with the Italian and German activist groups named "Last Generation:" *Ultima Generazione* and *Letzte Generation*. An online movement like this on the scale of #MeToo or #BlackLivesMatter—combined with ongoing disruptive protests and street action—could break the dam of denial and help provoke a great awakening.

Questions for Reflection and Discussion

- Are you in climate emergency mode? If not, are you ready to enter it? If not, why not?
- Do you talk about the climate emergency? Who do you avoid talking about it with? What are the barriers to talking about it more?

- What emergencies have you faced during your life? Do you remember the feeling during the moment, or extended period? What was it like and what were you like—focused and purposeful? Or overwhelmed and immobilized?
- Have you experienced the "movement mentality" and had a mission that was more important than yourself?
- What did your family members do during WWII? What are their memories and stories from that time?
- Have any of your family or friends been involved in ACT UP or in different social movements? What stories do they tell you about their experiences?

STEP FIVE: Join the Movement and Disrupt Normalcy

This morning I have climbed up this cable of the QEI Bridge of the M25 crossing the Thames... it's been really difficult. All we have to grip onto this cable is some rope slings. I'm willing to do this kind of thing, because... because I'm not willing to sit back and watch everything I love burn for the rest of my life. And that's what's happening....

So I've climbed up here today, and we will stay here until the governments make a meaningful statement that they will end licensing [for new oil drilling] and that they will initiate an emergency transition to renewable energy, funded by fossil fuel companies and the rich. We are out of time. Please join the resistance. You don't have to climb a bridge.... We are meeting daily at 11 am. Do an act of civil disobedience.

—Morgan Trow, Civil Engineer and activist with Just Stop Oil

ARE YOU READY? Have you faced climate truth and mourned your losses? Are you building emotional muscle—confronting your defenses and experiencing fear and other uncomfortable feelings? Have you realized that there is no individual solution to the climate emergency and that your self-interest is merged with the movement's interest? Are you convinced that nothing matters more than solving the climate catastrophe? If so, welcome to the team—the climate emergency movement. We demand that governments treat the climate like the existential emergency it is and initiate emergency mobilization to restore a safe climate. We are a movement channeling our grief into world-changing action. We are a movement made up of different people, campaigns, and organizations all over the world. We are nature, defending itself; we are on the team of the force on Earth that wants to live, and we draw tremendous strength from that.

The first edition of this book offered up a "big tent" vision of the climate movement. My strategic thinking has evolved. As the Executive Director of Climate Emergency Fund, I have spent hundreds of hours considering the most effective way to leverage our funds. I used to lose sleep over the question. We can't support a "good" approach—we don't have time for that. We need breakthroughs.

But now, I sleep soundly, as I have come to the conclusion that sustained escalating disruptive action is the fastest, most effective route to transformative change. Social movements are one of the most important driving forces of history—and disruption is a key driver for successful social movements. That is why Climate Emergency Fund exclusively funds groups who take part in disruptive protest—because we believe it is the fastest way to create transformative change. Both social science and history lead to this conclusion.

Thankfully, the climate movement has been growing more ambitious, disruptive, disciplined, and strategic.

Climate Movement 1.0 was dominated by Big Green, gradualism, coziness with business interests, and insider politics. We saw the Waxman-Markey climate bill fail in 2009, during the Obama administration, with this approach.[1]

Climate Movement 2.0 introduced a new paradigm—transformation at emergency speed, not gradual reform.

The shift toward strategic, nonviolent, heroic disruptions began in 2016 at Standing Rock, where the Indigenous Water Protectors showed the country what heroism looks like, withstanding months of abuse from police and private security forces while continuing their nonviolent, highly spiritual direct action of blocking the Dakota Access Pipeline.[2] Congresswoman Ocasio-Cortez, who has championed the Green New Deal and Climate Emergency frameworks in Congress, cites her time at Standing Rock as critical to her decision to run for Congress.

Standing Rock Lakota Nation member Floris White Bull describes Earth's vital life-force calling her to become a Water Protector:

I've been woken
by the spirit inside that
demanded I open my eyes
and see the world around me.
Seeing that my children's future
was in peril. See that my life couldn't
wait and slumber anymore. See that I was
honored to be among those who are awake.
To be alive at this point in time is to see the rising
of the Oceti Sakowin. To see the gathering of nations
and beyond that, the gathering of all races and all faiths.
Will you wake up and dream with us?
Will you join our dream. Will you join us?[3]

Climate Movement 2.0—propelled by new, confrontational grassroots groups like Extinction Rebellion, The Sunrise Movement, and the Fridays for Future school strikers—were all part of this wave, growing in power and influence during 2019 and early 2020.

COVID-19 presented a huge stumbling block to movement momentum. For example, Earth Day 2020 was on track to be the largest environmental demonstration in history. Instead, it was a livestream.

Thankfully, Climate Movement 3.0 is here! I have discussed some of these groups, such as Scientist Rebellion, Just Stop Oil, and New York Communities for Change in this book, but there are disruptive campaigns and groups popping up all the time. Climate Emergency Fund, in some ways, operates as a "venture philanthropy,"[4] funding very early-stage groups with relatively small grants that facilitate the coordination and recruitment of volunteer activists.

The primacy of disruptive protest is an inconvenient conclusion for me, personally. I find protests loud and overwhelming. I often feel out of place at them. My prior climate work was focused on creating a paradigm shift in language and drastically raising expectations of what we can accomplish if we enter emergency mode. I much prefer writing and thought leadership to protest. Disruptive protest is not my temperamental or aesthetic affinity; I support it based on strategic calculation.

If you, like me, you have trouble imagining yourself directly participating in a disruptive protest, getting arrested, or going to jail, that's okay. There are so many support roles that need to be filled, so many ways to use your skills and inclinations in support of the disruptive climate movement. Just don't confuse personal discomfort for strategic evaluation, something I see happen all too often. Nonviolent civil resistance is our best hope.

Analyzing the effects of activist movements leading up to recent US climate legislation, researchers at Giving Green calculate

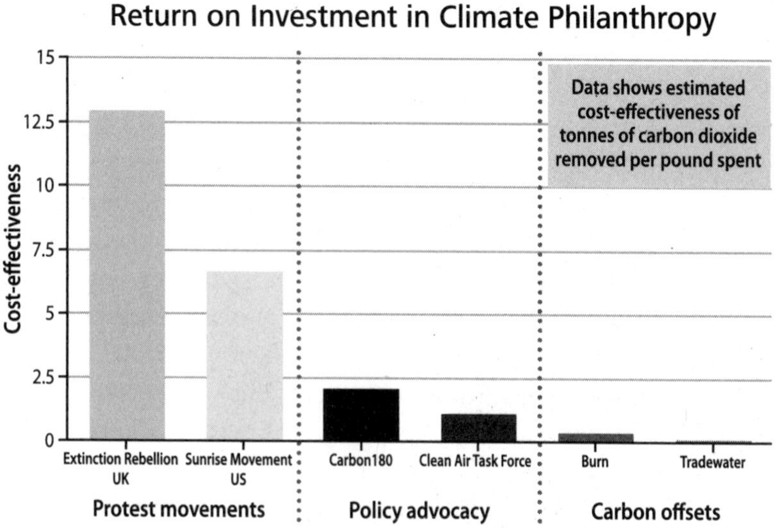

that, when political conditions are right, in the United States every "dollar spent on activism could remove more than 6 metric tons of CO_2e" (CO_2-equivalent) through galvanizing policy change, like the Inflation Reduction Act. The Social Change Lab used Giving Green's analysis and extended it to Extinction Rebellion, which in 2019 pushed the United Kingdom to pass a legally binding net zero emissions target, the first of any major economic power, in 2019 after weeks of disruptive protest. The amazing return on investment makes sense from a structural perspective—movement groups are powered by passionate and dedicated *volunteers*. Staff does exist, but they facilitate, coordinate, train, and support volunteer activists.

I had the joy of supporting Extinction Rebellion in their very early days. I was a strategic advisor to XR before they were known to the world, convincing them to take on the Declaration of Climate Emergency as a core demand of their bridge-blocking actions. And it worked. London declared a Climate Emergency just two weeks after their "Rebellion Day 1" bridge-building action. At that point, it had taken Climate Mobilization months of work to get the initial eight US declarations passed, to then have Bristol, London, and other UK cities pass it so quickly—this is power of disruptive action!

I spent a week in London with Extinction Rebellion in the lead-up to "Rebellion Day 1" in November 2018, supporting their fundraising and scaling. Working in XR's London office was one of my life's peak experiences; that office was buzzing with hope and energy—new people constantly showing up and saying, "How can I help?" It was a whirlwind of activity and, importantly, joy. Every two hours, people would end their calls, the lights would dim, and the all-volunteer group would have office-wide dance breaks.

I became Extinction Rebellion's first fundraiser. I had several years of fundraising experience under my belt from Climate Mobilization, and money seemed to be the last thing XR was thinking about. They were dedicated volunteers, focused like a laser on disruptive action. I added a Donate button to their website and held their first donor call.

While this was meaningful, nothing prepared me for the transformative experience of joining a bridge-blocking action at the end of my work with XR London.

I overcame my trepidation at participating in protests and linked arms with my comrades. The moment was exhilarating, a culmination of my mission, my grief, and my joy. I felt a sense of profound oneness with my fellow activists, an intense connection to the Earth, and deep commitment to our shared future.

I brought my experience back stateside, where I watched XR galvanize the global climate movement, raising ambitions and participation to new heights. I continued to raise funds for XR.

I was able to apply some of what I had learned in advising the visionary founders of Climate Emergency Fund (CEF), Aileen Getty, Rory Kennedy, and Trevor Neilson, who had been inspired by ACT UP and Extinction Rebellion and were creating funds to make grants that benefit the movement.

Both Getty and Kennedy come from famous families. Getty is an oil heiress, the granddaughter of J. Paul Getty. Kennedy is the youngest child of Ethel Kennedy and Senator Robert F. Kennedy, the younger brother of President John F. Kennedy. CEF's founders saw that the climate crisis required radical solutions, and put their considerable resources behind catalyzing them. They, along with Sarah Ezzy, Shannon O'Leary Joy, Geralyn Dreyfous, and Adam McKay, have committed to funding and supporting maximally effective action. In so doing, they have helped usher in the most powerful iteration of the climate movement yet.

Climate Emergency Fund's strategy is informed by history. In *This Is an Uprising*, Paul and Mark Engler lay out the history, theory, and shocking efficacy of social movements. As they describe, the first known act of civil disobedience was in 495 BC,[5] when the commoners of Rome rose up, demanding rights and the rule of law rather than rule by a capricious autocrat. First, the

Plebeian activists held public assemblies to spread their message and gain support. Then, they disrupted debt-collection hearings by yelling and making noise outside the hearing. They began to disobey laws en masse, initiated a general strike, and in their final act of escalation, they left Rome and set up camp at the Sacred Mount three miles outside the city. The Patricians had no choice but to capitulate to some of their demands immediately, and in coming decades, widespread "legal reform" was implemented; Roman citizens gained legal rights, and humanity leapt forward.[6]

Thousands of years later, these tactics somehow feel familiar. Climate protestors disrupted *The View's* conversation with Ted Cruz by yelling about the need for more climate coverage.[7] Occupy Wall Street set up an encampment. Strikes and union organizing have intensified since the pandemic upended labor markets around the world.[8]

While violent conflict seems to be a sad part of the human endowment, the ability to wage nonviolent conflict appears to be a core human capacity, as well. The Roman commoner uprising is one of over 1,200 nonviolent campaigns occurring across thousands of years and around the globe, cataloged in Swarthmore's Nonviolent Direct Action (NVDA) Database. Examples range from the Indian Independence struggle to the American abolitionists, the suffragettes to the campaign Indigenous Guatemalans led for rights, to the successful student-led campaign at Harvard to organize a workers union.

Nonviolence is a strategic imperative for movements, not just a moral one. Movement scholars Erica Chenoweth and Maria Stephan (2012) have found that nonviolent movements to overthrow authoritarian governments are more than twice as likely to succeed as violent movements, because "nonviolent campaigns facilitate the active participation of many more people than violent campaigns, thereby broadening the base of resistance and raising the costs to opponents of maintaining the status quo."

In the 1970s, scholar Gene Sharp identified 198 resistance tactics utilized by movements, including 19 forms of strikes and 9 types of boycotts. Tactics include public speeches, marches, distribution of literature against the government, mock funerals, various social types of boycotts, refusal to accept appointed officials, and the building of alternative social institutions.[9]

According to Sharp, nonviolence is an alternative to war or armed conflict, but it is still a type of combat:

> Nonviolent action is a means of combat, as is war. It involves the matching of forces and the waging of "battle," requires wise strategy and tactics and demands of its "soldiers" courage, discipline and sacrifice. This view... is diametrically opposed to the popular assumption that, at its strongest, nonviolent action relies on rational persuasion of the opponent, and more commonly it consists simply of passive submission.[10]

Nonviolent conflict has similarities to war. And yet it also has similarities to psychotherapy. Therapy helps an individual transform and heal, and movements help society transform and heal. Both are challenging processes powered by facing hard truths. Neither the therapist nor the activist believes the person or society they are trying to help change is irredeemable. Both believe in the possibility of transformative change. Dr. Martin Luther King Jr. spoke about how the nonviolent movement was healing a sick nation.[11] Climate activists are trying to wake up the public from the delusional trance of normalcy. Our protests may be controversial and inconvenient, but fundamentally, we are here to help. We are here out of love, hope, and responsibility.

Nonviolent direct action and disruptive protest create transformative impact, but it's indirect and can be hard to measure. Protests change public opinion and voting patterns and put relentless pressure on politicians.

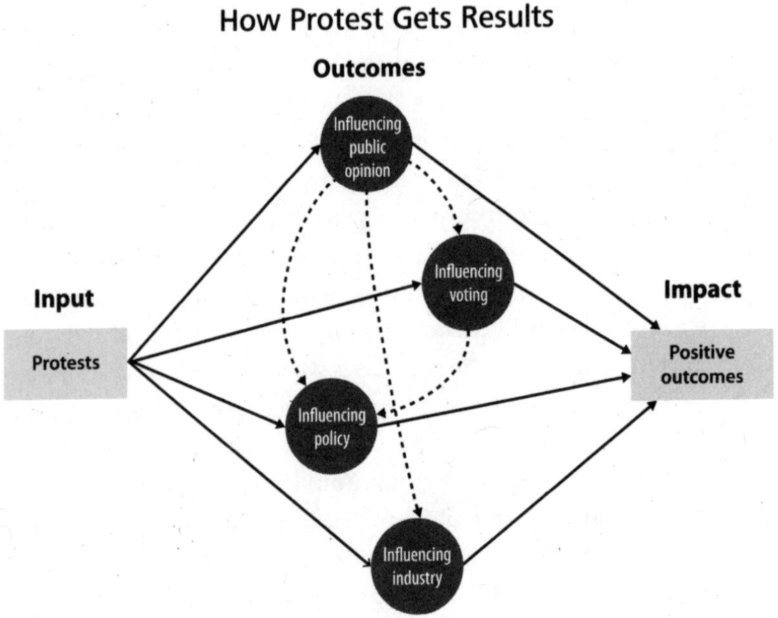

Illustration of how protests get results. CREDIT: JAMES OZDEN, SOCIAL CHANGE LAB.

Research examining thirty years of US electoral data shows that protest for "liberal" issues led to a greater share of votes for Democrats, while protest for "conservative" issues led to a greater share of votes for Republicans.[12] The author argues that protestors raise the "salience" of a certain issue, so when voters go to the polls, the issue ranks higher in their mind. The effect of salience on voting explains why larger protests with a higher intensity created a larger impact on election outcomes, and why disruptive protest was *not* found to create electoral "backlash."

We at Climate Emergency Fund are tremendously proud of the brave game-changing activists we support. But we are also aware of the immense untapped potential for a disruptive movement. We have seen small groups of dedicated brave grantees make a huge impact, such as through protesting Joe Manchin. Imagine what 10,000, 100,000, or even 1 million disruptive nonviolent protesters would accomplish. We could actually win.

A 2021 survey by the Yale Program on Climate Change Communication found that "millions of Americans [are] willing to support organizations and personally participate in nonviolent civil disobedience against corporate or government activities that make global warming worse." If the sample surveyed is representative of the population, it would mean that 8.6 million adults are "definitely" willing to personally participate in nonviolent civil disobedience![13] This is a staggering number. This is what we need to aim toward.

I hope I have convinced you to push your comfort zone and seriously consider using your skills, energy, and resources to support a disruptive protest group. You don't have to throw soup or get arrested. You can protest on the streets, or you can support the activists from your home office, like me.

But it's easier said than done. Figuring out *how* to join the movement in a good way for you is complicated. Finding your best role in the movement can be as complicated and multidimensional as choosing a career path. Only you can decide where you can be most effective. Further, people's roles in the movement change and grow all the time. I have had at least five different roles in the movement in various groups. To assess where you can most help the movement, I have four questions for you to consider:

- Your Body: What risks are you willing to take?
- Your Time: How many hours can you give to the movement?
- Your Skills: What special skills can you offer the movement?
- Your Wallet: Are you willing to give away money or fundraise?

Once you have self-reflected and evaluated these points, you will be ready to join the movement thoughtfully and effectively.

Your Body

Being willing to disrupt normalcy is like having a superpower. Even a small group of people who are willing to step up and take risks can have a huge impact. But it's not easy. Activists get hurt, get fined, and jailed. Some give their lives to the cause; globally, 1,700 environmental defenders have been murdered in the last ten years.

Acutely aware of this risk, Martin Luther King Jr. advised a group of movement leaders considering joining the nonviolent campaign in Birmingham, Alabama: "I think everybody here should consider very carefully and decide if he wants to be with this campaign.... I have to tell you that in my judgment, some of the people sitting here today will not come back alive from this campaign. And I want you to think about it."[14]

No one should be pressured to take part in arrestable actions—or any action. Everyone must assess their own risk tolerance. As someone who has never been arrested, I can assure you there are many ways to help. Please don't judge yourself or others, if this approach does not fit. People who are living paycheck to paycheck risk more than people with a financial cushion if they are fined, or arrested and miss work. Single parents or caregivers of an ill family member might not be able to be held overnight.

The often racist and brutal carceral system in the United States makes taking arrestable actions a nonstarter for many people of color in the US. Young activists of color have shared with me—and with others in movement spaces—that getting arrested often feels too risky. Some worry about how it would affect their families, when other family members have been wrongfully arrested or jailed.

With all of those caveats, I hope you do seriously consider taking nonviolent direct action and putting your body on the line. If you have racial or economic privilege, this is a great way to deploy it. If you don't have racial or economic privileges but are so passionate about the climate emergency and direct action, then this is a great way to deploy your bravery and passion as well. Everyone must make their own choice. Extinction Rebellion

NYC activist Mun Chong explained to *New York Magazine*, responding to the idea that the Extinction Rebellion is too white and middle class:

> I am from Malaysia, and we have no freedom of speech. I'd probably be dead or in jail right now doing what I do here if I were back home. So if the white middle-class folks have the privilege to get arrested and are willing, they should. No one should be holding back. Because we have everything to lose.[15]

Consider your risk tolerance:

- Are you willing to get arrested?
- Are you willing to put your body at risk?

Your Time

The best activists have assessed their schedules, determined the number of hours they can commit to the cause, and joined the movement in a capacity that matches their passion and their availability. They are flexible; they are responsive; they are reliable; and they take initiative. You've read this far, so I imagine you are a dedicated, thoughtful person that any organization would be lucky to have.

However, it may take some flexibility and persistence. It can be frustrating for new activists to feel their skills are not well used, but keep in mind that climate emergency organizations are usually small and budget-constrained. Onboarding and managing volunteers and activists can be time-intensive. Organizations can make great use of reliable volunteers and activists who are flexible enough to take on any task that needs doing. Consider your role as a supporting one, at least in the beginning. Work to support others in what they are already working on, get to know the lay of the land, and grow into a leadership role.

To determine your role in the climate emergency movement, ask yourself how many hours per week you can commit to climate work. Aim high. Although your first reaction might be that you don't have much time to devote, think again about that reflexive response. Might your claim to a lack of time be residual reluctance to take the climate crisis as an existential threat? You wouldn't be too busy to fight a raging fire, would you? Your time is valuable and is no doubt limited, but push yourself to devote at least seven hours a week to the climate emergency movement. While a volunteer with twenty-five hours a week can develop and manage an entire recruitment campaign, a volunteer devoting seven hours a week can help, for example by sending text messages to potential recruits.

Obviously, the more time, the better. As the director of a start-up organization that ran on volunteer time, I can't overstate the importance of volunteers who can dedicate twenty-five hours a week or more; it's just so much easier to coordinate people and get things done with people who are available throughout the day. Many volunteers and activists sacrifice comforts to give their time to a cause. I've seen activists leave high-powered jobs and move in with their parents to facilitate their round-the-clock climate work. If we are going to protect humanity and the natural world, some of us are going to need to go "all-in, for all life." Some of us must take responsibility for solving the climate crisis. Some of us must decide to sacrifice our ambitions and comforts. To determine if you are now or can be in this position, consider the following:

What is the greatest hourly and weekly commitment you can make?

- For six months? For one year? For three years?
- Could you reduce your hours at work or your time spent on other activities to allow for more climate work?
- Do you have savings or assets to cash in to support more climate work? (Or donate?)

- Do you have a partner who is willing to support your ambition to work full-time to save the world?
- Can you move in with family or roommates to cut your housing expenses?
- Can you ask the people who love and respect you to help support you to take a "climate year"?

Your Skills

Most people in the climate movement are in some kind of recruitment or activism role, which is great; they are the beating heart of the movement. But there are plenty of jobs for specialists as well. For example, research, social media, graphic design, and copyediting are just some of the functions that movements need and volunteers can fulfill. The time commitment will vary based on the skill in question and the level of coordination necessary. Below is a list of specialized tasks climate emergency organizations almost always need accomplished and would very often be happy to have a volunteer perform.

- Event planning
- Project management
- General administrative support
- Publicity and press support
- Graphic design; web and user-interface design
- Database design and management
- Tech support
- Social media
- Bookkeeping and accounting
- Legal advising
- Copyediting
- Cooking/catering
- Video and audio production
- Research
- Human resources and recruiting support
- Childcare

If you have these skills, reach out and offer your time to your climate emergency organization of choice.

Another way to use your skills is by supporting activists directly rather than through an organization. This requires less coordination and management. If you have professional skills to offer, even a few hours per week can be extremely helpful to organizers.

If you are a professional who offers clients services, consider making a portion of your work pro bono or available at a reduced fee for people in the climate emergency movement. For example, lawyers can represent activists who have been arrested; tax preparers could help activists with their personal tax returns; real-estate brokers can help organizers find situations for affordable co-housing. Or, you might be a professional who offers services that can help organizers avoid burnout. Mental health professionals can offer activists free individual or group therapy, so can massage therapists or acupuncturists could offer free or pay-what-you-can sessions. Whatever your area of expertise, consider how activists could benefit from it.

To identify the skills you can bring to volunteer work, answer the following questions:

- What skills do you have that organizers or organizations could benefit from?
- How many hours per week can you volunteer, and how flexibly?
- How will you effectively communicate to organizations or organizers that your skills are available?

Giving

Our culture tells us that building wealth is building security for ourselves and our families. Money is safety. Money is freedom. A friend told me that his only moral obligation in the face of the crisis was to "build wealth for my family." That view is utterly nonsensical in the age of ecological breakdown. We're told to save "for the future," but the ecological crisis is destroying that future.

Individuals, institutions, and governments all need to deploy the financial resources we have to protect humanity and restore a safe climate, now. This is what emergency mode means, not hoarding money for an imagined, denial-based future but rather mobilizing our resources to stop the emergency and restore safety. We cannot let civilization collapse while our money sits in banks or the stock market. Philanthropies, institutions, and individuals like you and me must give aggressively to organizations and candidates who support emergency climate mobilization.

When the mobilization arrives, the government should spend without limit to save as much life as possible. Until then, deploying your own financial resources now for the climate movement is one of the most impactful ways you can possibly use your money. In fact, supporting the movement financially is not only a moral duty, it is also the best way to achieve individual and familial security.

True security cannot coexist with the climate emergency and the sixth mass extinction of species. True security can only be achieved through the restoration of a safe climate and the natural world. Therefore, not only do I suggest that you give money to the climate emergency movement, I also suggest that you hold *all* organizations and politicians to a climate emergency standard for giving.

This means that if you are a contributor to the National Resources Defense Council, the Environmental Defense Fund, or even an "unrelated" organization like an art museum or a university, you tell fundraisers and others that you will support them only if they start treating climate like the emergency it is through their spending, communications, and policy advocacies. Hold this line for politicians, too. If they aren't making the climate emergency their top priority and advocating WWII-scale climate mobilization that eliminates emissions in ten years or less, stop supporting them—and tell them why.

To better prepare to contribute your money to the climate emergency movement, ask yourself if you can imagine making a financial commitment that feels commensurate with the scale and urgency of the climate crisis itself. If the answer is yes—and it must be—then ask yourself:

- How much money do you have saved?
- What is your net worth?
- What is your annual income?
- How much do you spend a year, and what are your biggest expenses?
- What one-time donation could you make that would push your capacity for giving?
- What maximum monthly contribution could you commit to for one year?
- What lifestyle sacrifices could you make to donate more money to climate change mobilization?
- What other tangible or intangible resources can you commit (a meeting space, a vehicle, etc.)?
- What climate emergency organizations are you most impressed by? How can your money help their efforts?

There's an obvious tension to call out between giving your time and giving your money. Should you stick with a high-paying job or career path and donate every extra cent to the movement? Or should you live on the bare minimum and dedicate all your time and skills to the cause? The reality is that the movement needs both kinds of people. What's important is that you are a person who has woken up and is all-in to rescue human life and the future.

Fundraising

This volunteer task is very high leverage but massively unpopular. Just as we're conditioned by the capitalist system to earn as much money as we can and pile it up, we are conditioned that it's shameful to ask for help.

As I have gone through my climate journey, trying to make the most effective contributions to the climate movement I can, I have done more and more fundraising. When I started, my hands would shake in meetings. To hype myself up, I would tell myself before fundraising calls: *You are doing something brave.*

Now, it's just another day at the home office. I have come to realize that asking people for money for climate activism isn't some vulgar, grasping financialized arrangement. It's a critical part of the mission, and those who are willing to do it are soldiers for the cause. More and more philanthropists are reckoning with climate truth, seeing that normal channels have failed, and are looking to support bold climate activism. With Climate Emergency Fund, I help them direct their funds to the highest-impact groups.

In her classic book, *Fundraising for Social Change*, Kim Klein discusses this process of "getting over the fear of asking" by recognizing that the cause is more important than the fear.[16] Fundraising for the climate movement is a challenging but extremely high-impact part of our fight for civilization and the natural world.

Klein encourages fundraisers to view soliciting donors as offering them an opportunity to be a part of something meaningful. In this case, supporting the climate movement isn't just charity—it's our only hope. Despite the fantasies of space-exploring, luxury-bunker-building billionaires, there is no place to hide from the climate crisis. We need to stand up and protect our shared home. How could money be put to better use?

Some people are under the illusion that social movement organizations shouldn't need money. They think these organizations should operate exclusively with volunteers and miracles. This is a beautiful concept—and it sometimes works for a limited period of time, as it did during Occupy Wall Street.

But to sustain themselves through months or years, social movement organizations need people who are available for meetings during the workweek. They need people who will, consistently and diligently, do the things no one else wants to do, like handling the legalities, answering emails, dealing with difficult people and situations, and keeping the organization financially solvent. Core full-time movement staff doesn't need to be paid a market rate—but they do need a living wage. Then there are other expenses, office space, travel, and printing, to name a few. Simply put, organizations need money!

For organizations that are considered "radical" and are outside the comfort zone of traditional philanthropic funders, securing funding can be extremely challenging.[17] This is why almost any organization will say, "Yes, please!" to someone offering to work as a fundraising volunteer, especially if that person can fundraise from within their own networks.

Fundraising can take many forms. It might mean inviting friends and family over for a discussion about the climate crisis and, at the discussion's end, asking for donations to your favored organization. It might mean asking for money through your social media presence. Facebook, for example, has a platform where you can fundraise for nonprofit organizations. It might mean that you pass the hat at work or email ten friends to tell them you are making a donation and asking them to match it. Climate emergency organizations might also ask fundraising volunteers to call past donors or people who are supporters but not donors and ask them for money, or to do background research on donors, or help prepare grant applications.

To assess your fundraising preparedness, answer the following questions:

- How wealthy and philanthropic is your professional and social network? Do they regularly donate to social causes?
- Have you ever asked someone to donate money to a cause? How comfortable or uncomfortable were you?
- What about fundraising is appealing or repellent?
- Are you willing to try overcoming social awkwardness and fears about asking?

Getting Started

We must respond to the climate emergency not only with our heads and hearts: We have a moral obligation to respond with dedicated and fierce action. Reflect on where you fit in and consider your "highest and best use," but don't get stuck in analysis paralysis.

Determining your place in the movement is an ongoing process. Have a conversation with someone about the climate emergency today and every day. Make a financial contribution today, and then start thinking about your next contribution. Attend a meeting this week, and grow your role from there. Above all, remember that your actions are contagious and can have a huge impact on others. By breaking the silence and entering emergency mode, you will lead your friends, family members, coworkers, and neighbors to join you.

Questions for Reflection and Discussion

- How many hours a week will you commit to the climate emergency movement?
- What are the biggest barriers to dedicating even more time?
- What do you think your highest and best use might be to the movement?
- What potential roles seem the most appealing to you? Which roles do you want to avoid?

CONCLUSION: ALL-IN FOR ALL LIFE

For all people, all species, and all generations.

—Philip Sutton

THIS IS MEANT TO BE A PRACTICAL BOOK. It is intended to support your becoming a more effective change agent in initiating the great awakening to save humanity and life on Earth. Thank you for your courage in confronting the pain of the climate emergency and for seriously considering how to maximize your contribution.

Dedicating yourself to canceling the apocalypse and protecting all life is the biggest challenge you will ever take on. When you set your ambitions on saving the world, there is always more to do. Your career may suffer. You will experience conflict as well as deep camaraderie with your activist comrades. You may even go to jail. Do it anyway. Even though you don't have every skill, even though you have personal faults, you are good enough to do this work. It is your birthright.

Neoliberalism has told us that we are self-interested consumers who are cynical and passive about politics. But that's not true. The crisis we're facing demands that we define ourselves differently—we must each see ourselves as someone who will take personal responsibility for solving the global climate and ecological crises, as someone who is willing to dedicate themselves to inspiring others to face climate truth, and as someone who can help initiate full-scale mobilization. The crisis demands we who hear the call to heroism heed it. The crisis requires that we believe in our fundamental interconnection—to each other and to all life. It demands that we stop lying to ourselves and others, and that we stop using defenses and euphemisms. It requires that we are ready to be transformed by the truth and are able to lead others in transformation.

In 1943, while the home front mobilization was underway, American psychologist Abraham Maslow developed his famous hierarchy of needs.[1] The pyramid offers an approximate visual representation of his concept, but was not created by Maslow himself.

Rather than studying mental illness, Maslow investigated how humans flourish. He believed that people could achieve their highest potential only by first addressing their basic needs. Before we can be free to meet our needs for safety, esteem, love, and belonging, we must first meet our needs for food, clothing, and shelter. If we do not satisfy our basic needs, such as our need to eat, we will spend all of our time and energy looking for food.

However, once we have satisfied our basic needs, we can expect new discontentedness and restlessness to develop unless we

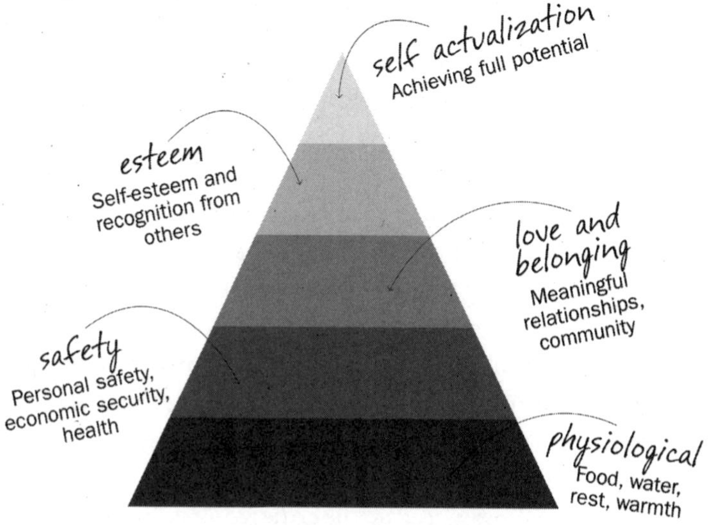

Visual Approximation of Maslow's Hierarchy of Needs

CREDIT: JEFFREY SAKS.

pursue our highest calling and pursue our true potential. If they are to be happy, a scholar must study, a carpenter must build, and a musician must make music. What a person *can* be, they *must* be. Maslow refers to this need as self-actualization.

Self-actualization may sound self-centered. It is anything but. It is through meeting our basic needs and then tapping into our truest selves and potential that we are able to serve something greater than ourselves. Maslow studied self-actualized people and wrote about their characteristics. Far from characterizing them as self-absorbed, he described them as "problem-centered." Self-actualized people have a mission in life, a task to fulfill, a problem outside of themselves that enlists much of their energies. These aren't necessarily tasks that they would prefer or choose for themselves. They may be tasks that they feel are their responsibilities, duties, or obligations.[2] Toward the end of his life, Maslow described the need for "self-transcendence." He wrote: "The good of other people must be invoked, as well as the good for oneself."[3]

Chloe Bertini, an Italian activist with Ultima Generazione, is a wonderful example. A passionate professional dancer, Chloe gave up her spot in the *Lion King* tour in order to pursue disruptive activism because it allows her to live in truth and be "in coherence with myself." Chloe stepped back from a creative, athletic, joyful, successful career in order to take up an even higher calling.

In this time of acute ecological crisis, being creatively, passionately, and humbly engaged in the climate emergency movement is the definition of self-actualization. We didn't choose this emergency; we would probably rather be doing something else. But here we are, in personal and collective danger. The climate emergency chose us, and we must devote our vast potential to reversing it.

It is an honor and a privilege to walk the path of the climate activist, but it's a privilege that is unavailable to many. Across the country and the globe, many people already lack food, security, and shelter. Wealth inequality, combined with the climate emergency, makes life extremely difficult for many. As the Earth continues to warm, people who now claim only precarious security will likely face tremendous challenges. People who are in danger of being evicted or going hungry cannot be expected to prioritize global cataclysm ahead of their daily existential risk—although frontline and Indigenous activists from Ecuador to United States to Congo are managing to do both at the same time.

So few people have the freedom and ability to choose this heroic path and go all-in for all life that if you can choose it, then you must. And, of course, remember that there are many legal and quieter ways to support the disruptive climate movement.

However you get involved, this path is richly rewarding. You will feel lit up and renewed by the mission. You will be awed by the immensity and beauty of all life. You will be grateful and proud to be in its service. You will have and you will feel connection and belonging with your fellow activists, protectors of humanity and the natural world.

When we undertake the work of transformation, we don't merely become "better." We become something entirely new. Climate activists sometimes draw inspiration from one of nature's most dramatic transformational processes—a caterpillar becoming a butterfly. As environmentalist Kim Polman writes in *Imaginal Cells: Visions of Transformation:*

> After a period of ravenous consumption, the caterpillar forms a chrysalis from which it will dissolve itself into an organic stew, where dormant "imaginal cells" hold the vision of the new structure. At first, these imaginal cells operate independently, as single-cell organisms, and are attacked by the caterpillar's immune system, which views them as a threat. But soon, these new cells regroup, multiply,

> and connect with each other. They then form clusters and begin resonating at the same frequency. Finally, they reach a tipping point and consolidate to become a new multicellular organism, the beautiful butterfly.[4]

The caterpillar consumes voraciously, but its imaginal cells lead to its transformation into a butterfly—a pollinator and a powerful contributor to the health of the entire ecosystem. What the caterpillar takes, the butterfly gives back.

Humanity, especially those of us in the United States and other affluent countries, has consumed voraciously. We are capable of so much more. It's time to transform and give back. We need a society full of healers and protectors. It's time to shift from an economy based on destruction to one based on regeneration and care.

Now is our time to become humanity's imaginal cells and lead the transformation. We have been isolated and dormant for too long, but now we are organizing, growing, and building a movement. We know we will face the parts of the old system that are afraid, the parts that view change as a threat. But we will withstand attacks because we know our work to lead society to the tipping points is necessary for transformation.

It is not a given that we will successfully transform, that this movement will win with enough time to avert civilization's collapse, but it is our only hope. We must join together to do everything we can to initiate emergency mobilization as quickly as possible. We must turn our pain into action and take personal responsibility for protecting humanity and the living world. We must take our rightful place in the disruptive climate movement. We must become heroes.

Onward!

ENDNOTES

Preface: A New Age of Heroes

1. Caroline Hickman et al., "Climate Anxiety in Children and Young People and Their Beliefs About Government Responses to Climate Change: A Global Survey," *The Lancet Planetary Health* 5, 12 (2021).
2. FAO, IFAD, UNICEF, WFP and WHO, *The State of Food Security and Nutrition in the World 2022: Repurposing Food and Agricultural Policies to Make Healthy Diets More Affordable* (Rome, FAO, 2022).

Introduction

1. Jared Keller, "The U.S. Suicide Rate Is at Its Highest in Half a Century," *Pacific Standard*, December 4, 2018 (accessed July 23, 2019).
2. "Mental Health Information: Statistics: Suicide," National Institute of Mental Health, U.S. Department of Health and Human Services, June 2022 (accessed November 28, 2022).
3. National Safety Council: Injury Facts, "Preventable Deaths" (accessed July 23, 2019).
4. Thomas Moore and Donald Mattison, "Adult Utilization of Psychiatric Drugs and Differences by Sex, Age, and Race," *JAMA Internal Medicine* 177 (February 2017): 274–75 (accessed July 23, 2019).
5. Ibid.
6. Nielsen, "The Total Audience Report: Q1 2016," *Nielsen* (accessed July 23, 2019).
7. Genevieve Guenther, "Who Is the 'We' in 'We Are Causing Climate Change'?" *Grist*, October 13, 2018 (accessed July 23, 2019).
8. Erich Fromm, "Love of Death and Love of Life," *The Heart of Man: Its Genius for Good and Evil*, ed. Ruth Nanda Anshen (New York: Harper, 1964).
9. Doris Kearns Goodwin, *No Ordinary Time: Franklin & Eleanor Roosevelt: The Home Front in World War II*, 1st ed. (Simon & Schuster, 2008).
10. Oscar Wilde, *The Soul of Man Under Socialism* (Elsinor Verlag, 2021).
11. Philip Sutton, *Striking Targets: Matching Climate Goals with Climate Reality*, 2005.
12. Naomi Klein, *This Changes Everything: Capitalism vs. The Climate* (New York: Simon & Schuster, 2014).

13. D.J. Peterson, *Troubled Lands: The Legacy of Soviet Environmental Destruction.* (Boulder: Westview Press, 1993).
14. Ibid., 36; Global Carbon Project, "CO2 Emissions," Global Carbon Atlas (accessed November 12, 2019).
15. Norwegian Ministry of Trade, Industry and Fisheries, *The State Ownership Report 2015* (Norway: Norwegian Ministry of Trade, Industry and Fisheries, 2015), www.equinor.com/content/dam/statoil/documents/the-state-ownership-report-2015.pdf (accessed August 18, 2019).
16. Edward O. Wilson, *Half-Earth: Our Planet's Fight for Life* (New York: Liveright, 2016).
17. Pope Francis (Jorge Mario Bergoglio), *Encyclical Letter Laudato Sí' of the Holy Father Frances on the Care of Our Common Home,* May 24, 2015, http://w2.vatican.va/content/francesco/en/encyclicals/documents/papa-francesco_20150524_enciclica-laudato-si.html (accessed October 23, 2019).
18. Oxford University Press. "Oxford Word of the Year 2019," *Oxford Languages.*
19. Katharine Wilkinson and Britt Wray, "7 Resources to Help Manage Climate Anxiety," *Time,* November 3, 2021 (accessed November 28, 2022).
20. Albert Camus, *Lyrical and Critical Essays* (Knopf Doubleday Publishing Group, 2012).
21. Damien Gayle and Rob Davies, "No 10 Condemns 'Guerrilla Tactics' as Just Stop Oil Activists Block Fuel Depots," *The Guardian,* April 11, 2022 (accessed November 28, 2022).
22. Damien Gayle, "Just Stop Oil Activists Throw Soup at Van Gogh's Sunflowers," *The Guardian,* October 14, 2022 (accessed November 28, 2022).
23. Jason Duaine Hahn, "2022 Tour De France Disrupted by Climate Activists Blocking Course During Stage 10," *People,* July 12, 2022 (accessed November 28, 2022).

Step One: Face Climate Truth

1. Benjamin Chapman, Kevin Fiscella, Ichiro Kawachi, Paul Duberstein, and Peter Muennig, "Emotional Suppression and Mortality Risk Over a 12-Year Follow-Up," *Journal of Psychosomatic Research* 75, 4 (2013): 381–85.
2. David Wallace-Wells, *The Uninhabitable Earth: Life After Warming* (New York: Penguin Random House, 2019).
3. Megan Darby, "Meet the Woman Who First Identified the Greenhouse Effect," *Climate Change News,* February 9, 2016 (accessed July 24, 2019).

4. Marit-Solveig Seidenkrantz, "80 Years Since the First Calculations Showed That the Earth Was Warming Due to Rising Greenhouse Gas Emissions," *Science X*, June 5, 2018, phys.org (accessed July 24, 2019).
5. Neela Banerjee, Lisa Song, and David Hasemyer, "Exxon's Own Research Confirmed Fossil Fuels' Role in Global Warming Decades Ago," *Inside Climate News*, September 16, 2015 (accessed July 24, 2019).
6. Naomi Oreskes and Erik M. Conway, *Merchants of Doubt: How a Handful of Scientists Obscured the Truth on Issues from Tobacco Smoke to Climate Change* (Bloomsbury Publishing, 2011).
7. Katie Worth, "Climate Change Skeptic Group Seeks to Influence 200,000 Teachers," *Frontline, PBS.org* (accessed October 23, 2019).
8. "Eco-Bot.Net: Guide to Climate Disinfo." *Eco-Bot.Net* (accessed November 28, 2022).
9. Museum Exhibits, university departments, and advertising in the *New York Times*.
10. Robert J. Brulle, "Institutionalizing Delay: Foundation Funding and the Creation of U.S. *Climate Change Counter-Movement Organizations," Climatic Change: An Interdisciplinary, International Journal Devoted to the Description, Causes and Implications of Climatic Change* 122, 4 (2014): 681–94.
11. Robert J. Brulle, "The Climate Lobby: A Sectoral Analysis of Lobbying Spending on Climate Change in the USA, 2000 to 2016," *Climatic Change: An Interdisciplinary, International Journal Devoted to the Description, Causes and Implications of Climatic Change* 149, 3–4 (2018): 289–303.
12. Andy Rowell, "Fossil Fuel Industry Has Spent Nearly $2 Billion on Lobbying to Kill Climate Laws," *Oil Change International,* July 20, 2018, https://priceofoil.org.
13. Jonathan Chait, "Why Are the Republicans the Only Climate-Science Denying Party in the World?" *New York Magazine,* September 27, 2015 (accessed July 25, 2019).
14. "Leaders: A Greener Bush," *The Economist,* February 15, 2003 (accessed August 18, 2019).
15. Valerie Richardson, "Obama Takes Credit for U.S. Oil-and-Gas Boom: 'That Was Me, People,'" *The Washington Times,* November 28, 2018 (accessed August 18, 2019).
16. Douglass Rushkoff, *Survival of the Richest: Escape Fantasies of Tech Billionaires* (WW Norton, 2022).
17. Glenn Scherer, "Climate Science Predictions Prove Too Conservative," *Scientific American,* December 6, 2012 (accessed on July 26, 2019).

18. David Spratt and Ian Dunlop, *What Lies Beneath: The Understatement of Existential Climate Risk* (Melbourne, Australia: Breakthrough, 2018).
19. Michael Mann, Susan Hassol, and Tom Tole, "Doomsday Scenarios Are as Harmful as Climate Change Denial," *The Washington Post*, July 12, 2017 (accessed July 26, 2019).
20. Anand Giridharadas, *Winners Take All: The Elite Charade of Changing the World* (New York: Knopf, 2018).
21. Michael Tobis and Stephen Ban, "OK, Getting Serious Again," *Only In It for the Gold* (blog), Planet 3, January 14, 2010 (accessed July 29, 2019).
22. https://insideclimatenews.org/news/16092022/pakistan-flood-displacement/
23. https://www.who.int/europe/news/item/07-11-2022-statement---climate-change-is-already-killing-us--but-strong-action-now-can-prevent-more-deaths
24. Andrew Freedman, "Hurricane Ian May Have Been Florida's Costliest Storm, Estimate Suggests," *Axios*, October 7, 2022 (accessed November 28, 2022).
25. Reality Check and Visual Journalism, "China, Europe, US Drought: Is 2022 the Driest Year Recorded?" *BBC News*, September 17, 2022 (accessed November 28, 2022).
26. United Nations Climate Change, "Climate Plans Remain Insufficient: More Ambitious Action Needed Now," *Unfccc.int*, October 26, 2022.
27. David Spratt and Ian Dunlop, *Existential Climate Risk: A Scenario Approach* (Melbourne, Australia: Breakthrough, 2019).
28. IPCC, *Global Warming of 1.5°C: An IPCC Special Report on the Impacts of Global Warming of 1.5°C Above Pre-Industrial Levels and Related Global Greenhouse Gas Emission Pathways*, eds. V. Masson-Delmotte et al., www.ipcc.ch (accessed July 26, 2019).
29. "Planetary Boundaries Research," *Stockholm Resilience* Centre (accessed July 29, 2019).
30. "The Extinction Crisis," *Center for Biological Diversity* (accessed August 10, 2019).
31. *Living Planet Report 2018: Aiming Higher*, M. Grooten and R.E.A. Almond, eds., (Switzerland: WWF, 2018).
32. Caspar Hallman et al., "More than 75 Percent Decline Over 27 Years in Total Flying Insect Biomass in Protected Areas," *PloS ONE* 12, 10 (2017).
33. Cheryl Schultz, Leone Brown, Emma Pelton, and Elizabeth Crone, "Citizen Science Monitoring Demonstrates Dramatic Declines of

Monarch Butterflies in Western North America," *Biological Conservation* 214 (2017): 343–46.

34. Paul Gilding, *The Great Disruption* (London: Bloomsbury Press, 2011).
35. D. Lin et al., "Ecological Footprint Accounting for Countries: Updates and Results of the National Footprint Accounts, 2012–2018," *Resources* 7 (2018): 58.
36. Michael Borucke et al., "Accounting for Demand and Supply of the Biosphere's Regenerative Capacity: The National Footprint Accounts' Underlying Methodology and Framework," *Ecological Indicators* 24 (January 2013): 518–33.
37. Antonio Guterres and Patricia Espinosa, "Carbon Inequality in 2030," *OXFAM*, November, 2021 (accessed November 27, 2022).
38. Greenpeace. *Fossil Fuel Racism*, April 13, 2021 (accessed November 27, 2022).
39. U.S. Department of Agriculture, "Key Statistics & Graphics," *USDA ERS*, October 17, 2022.
40. Bill McKibben, *Eaarth: Making a Life on a Tough New Planet* (New York: Times Books, 2010)
41. Ibid., 2.
42. Ibid., 5.
43. John M. Darley and Bibb Latané, "Bystander Intervention in Emergencies: Diffusion of Responsibility," *Journal of Personality and Social Psychology* 8, 4 (1968): 377–83.
44. Bibb Latané and John M. Darley, *Group Inhibition of Bystander Intervention in Emergencies* (Emmitsburg, MD: National Emergency Training Center, 1968).
45. D. Nilsson and A. Johansson, "Social Influence During the Initial Phase of a Fire Evacuation: Analysis of Evacuation Experiments in a Cinema Theatre," *Fire Safety Journal* 44 (2008): 71–79.
46. Robert B. Cialdini, *Influence: Science and Practice*, 5th ed. (Harlow, Essex: Pearson, 2008).
47. Doris Goodwin, "The Way We Won: America's Economic Breakthrough During WWII," *American Prospect* (Fall 1992) (accessed July 29, 2019).
48. "Table 3.1," *Historical Tables: Budget of the United States Government Fiscal Year 2011*, White House Office of Management and Budget (accessed August 18, 2019).
49. Price V. Fishback and Joseph Cullen, *Did Big Government's Largesse Help the Locals? The Implications of WWII Spending for Local Economic*

Activity, 1939–1958 (Cambridge: National Bureau of Economic Research, 2006).

50. Doris Kearns Goodwin, *No Ordinary Time: Franklin & Eleanor Roosevelt: The Home Front in WW II* (New York: Simon & Schuster, 1994); David Kaiser, *No End Save Victory: How FDR Led the Nation into War* (Philadelphia: Basic Books, 2014); Arthur Herman, *Freedom's Forge: How American Business Produced Victory in World War II* (New York: Random House, 2013).
51. M.M. Eboch, *Native American Code Talkers* (Minneapolis, MN: ABDO, 2015).
52. Doris Kearns Goodwin, *No Ordinary Time*; David Kaiser, *No End Save Victory*; Arthur Herman, Freedom's Forge, 332.
53. Thomas Bassett, "Reaping on the Margins: A Century of Community Gardening in America," *Landscape* 25, 2 (1981): 1–8.
54. Lester Brown, *Plan B 4.0: Mobilizing to Save Civilization* (New York: WW Norton, 2009): 260.
55. Maury Klein, *A Call to Arms: Mobilizing America for World War II* (New York: Bloomsbury Publishing, 2015).
56. Hugh Rockoff, *Keep on Scrapping the Salvage Drives of World War II* (Cambridge, MA: National Bureau of Economic Research, 2007), http://papers.nber.org/papers/w13418.
57. National Science Foundation, *The National Science Foundation: A Brief History* (Washington, D.C.: ERIC Clearinghouse, 1988).
58. Andrew Edmund Kersten, *Race, Jobs, and the War: The FEPC in the Midwest, 1941–46* (Urbana: University of Illinois Press, 2007).
59. Ronald T. Takaki, *Double Victory: A Multicultural History of America in World War II* (Boston: Little, Brown and Company, 2001).
60. Kearns Goodwin, *No Ordinary Time*, 555.
61. Ibid., 769.
62. Ibid., 628.
63. Lydia Saad, "Gallup Vault: A Country Unified After Pearl Harbor," *Gallup*, December 5, 2016 (accessed August 18, 2019).
64. Ibid.
65. David Kaiser, *No End Save Victory.*
66. Robert J. Shiller, "Once Cut, Corporate Income Taxes Are Hard to Restore, *New York Times*, June 22, 2018
67. Kearns Goodwin, *No Ordinary Time*: 356–59.
68. Jeff Tollefson, "Covid Curbed Carbon Emissions in 2020, but Not by Much," *Nature News*, January 15, 2021 (accessed November 28, 2022).

69. Douglas Keay, "Interview for *Women's Own,*" Margaret Thatcher Foundation, October 16, 1984
70. John Duberstein, *A Velvet Revolution: Václav Havel and the Fall of Communism* (Greensboro, North Carolina: Morgan Reynolds, 2006).
71. Václav Havel, *The Power of the Powerless: Citizens Against the State in Central Eastern Europe,* trans. John Keane (Abingdon: Routledge, 1985).

Step Two: Welcome Fear, Grief, and Other Painful Feelings

1. Kerry Kelly Novick and Jack Novick, *Emotional Muscle: Strong Parents, Strong Children* (Xlibris: 2010).
2. Kristen Neff, *Self-Compassion: The Proven Power of Being Kind to Yourself* (New York: HarperCollins, 2011).
3. Jean Carlomusto, dir., *Larry Kramer: In Love & Anger,* 2015.
4. "RAIN: Recognize, Allow, Investigate, Nurture," *Tara Brach* (website) (accessed August 19, 2019).
5. Jaak Panksepp, "Affective Neuroscience of the Emotional BrainMind: Evolutionary Perspectives and Implications for Understanding Depression," *Dialogues in Clinical Neuroscience* 12, 4 (December 2010): 533–45.
6. Thierry Steimer, "The Biology of Fear- and Anxiety-Related Behaviors," *Dialogues in Clinical Neuroscience* 4, 3 (September 2002): 231–49.
7. Joanna Macy and Chris Johnstone, *Active Hope: How to Face the Mess We're in Without Going Crazy* (California: New World Library, 2012): 71.
8. Ibid.
9. DARA and the *Climate Vulnerable Forum, Climate Vulnerability Monitor, 2nd Edition: A Guide to the Cold Calculus of a Hot Planet* (Madrid: Fundación DARA Internacional, 2012).
10. WWF, *Living Planet Report 2016: Risk and Resilience in a New Era* (WWF International, Gland, Switzerland, 2016).
11. Mary Annaïse Heglar, "The Big Lie We're Told About Climate Change Is That It's Our Fault," *Vox,* November 27, 2018 (accessed July 31, 2019).
12. J. Rottenberg, F.H. Wilhelm, J.J. Gross, I.H. Gotlib, "Vagal Rebound During Resolution of Tearful Crying Among Depressed and Nondepressed Individuals," *Psychophysiology* 40 (2003):1–6.
13. William H. Frey and Muriel Langseth. *Crying: The Mystery of Tears* (Minneapolis, MN: Winston Press, 1985).
14. N. van Leeuwen, E.R. Bossema, H. van Middendorp, A.A. Kruize, H. Bootsma, J.W.J. Bijlsma, and R. Geenen, "Dealing with Emotions When the Ability to Cry Is Hampered: Emotion Processing and

Regulation in Patients with Primary Sjögren's Syndrome," *Clinical and Experimental Rheumatology* 30 (4) (2012): 492–98.

15. Asmir Gračanin, Lauren Bylsma, and Ad J.J.M. Vingerhoets, "Is Crying a Self-Soothing Behavior?" *Frontiers in Psychology* 5 (May 2014).
16. A.J.J.M. Vingerhoets, N. van de Ven, and Y. van der Velden, "The Social Impact of Emotional Tears," *Motivation and Emotion* 40 (3) (2016): 455–63.
17. Carolyn Baker and Guy R. McPherson, *Extinction Dialogues: How to Live with Death in Mind* (Next Revelation Press, 2014).
18. Carolyn Baker, *Collapsing Consciously: Transformative Truths for Turbulent Times* (Berkeley, CA: North Atlantic Books, 2013).
19. Joanna Macy and Chris Johnstone, *Active Hope*: 75.
20. Ibid.

Step Three: Reimagine Your Life Story

1. Margaret Klein, "The Trauma of a Romantic Partner's Psychotic Episode: An Emerging Clinical Picture," PhD diss, Adelphi University, 2014.
2. Henry Wadsworth, "The Rainy Day," *Longfellow*, www.hwlongfellow.org/poems_poem.php?pid=39.
3. Jeremy Lent, *The Web of Meaning: Integrating Science and Traditional Wisdom to Find Our Place in the Universe* (Gabriola Island, BC: New Society Publishers, 2021).
4. Edward O. Wilson, *The Social Conquest of Earth* (New York: Liveright, 2012): 241.
5. Martin Luther King Jr., *Letter from a Birmingham Jail*, August 1963 (accessed August 1, 2019).

Step Four: Enter Emergency Mode

1. David Gelles, "Billionaire No More: Patagonia Founder Gives Away the Company," *New York Times*, September 14, 2022 (accessed November 28, 2022).
2. Maury Klein, *A Call to Arms: Mobilizing America for World War II* (New York: Bloomsbury Publishing, 2015); Doris Goodwin, "The Way We Won: America's Economic Breakthrough During WWII," *American Prospect* (Fall 1992) (accessed July 29, 2019).
3. David Kaiser, *No End Save Victory: How FDR Led the Nation into War* (New York: Basic Books, 2015).
4. Daniel C. Ehlman, Ellen Yard, Deborah M. Stone, Christopher M. Jones, and Karin A. Mack, "Changes in Suicide Rates: United States, 2019 and 2020," *Morbidity and Mortality Weekly Report* 71, 8 (2022): 306–12.

5. Roger S. McIntyre et al., "Suicide Reduction in Canada During the COVID-19 Pandemic: Lessons Informing National Prevention Strategies for Suicide Reduction," *Journal of the Royal Society of Medicine* 114, 10 (2021): 473–79.
6. Sebastian Junger, *Tribe: On Homecoming and Belonging* (Twelve, 2016).
7. Rebecca Solnit, *A Paradise Built in Hell: The Extraordinary Communities That Arise in Disaster* (Penguin Publishing Group. Kindle Edition).
8. David Spratt and Philip Sutton, *Climate Code Red: The Case for Emergency Action* (Carlton North, Victoria, Australia: Scribe Publications, 2009).
9. Joseph Stiglitz, "The Climate Crisis Is Our Third World War: It Needs a Bold Response," *The Guardian*, June 4, 2019 (accessed August 18, 2019).
10. Thomas Friedman, *Hot, Flat, and Crowded: Why We Need a Green Revolution—and How It Can Renew America* (New York: Farrar, Straus and Giroux, 2008).
11. Naomi Klein, *No Is Not Enough: Resisting Trump's Shock Politics and Winning the World We Need* (Chicago: Haymarket Books, 2017).
12. Bill McKibben, "A World at War," *The New Republic*, August 15, 2016 (accessed August 18, 2019).
13. Bracken Hendricks, Rhiana Gunn-Wright, and Sam Ricketts, "The Greatest Mobilization Since WWII," *Democracy: A Journal of Ideas*, March 10, 2020.
14. Harvard Thinks Big 2010, *Daniel Gilbert: Global Warming and Psychology*, Vimeo video (accessed August 12, 2019).
15. Jean Carlomusto, dir., *Larry Kramer in Love and Anger* (HBO, 2015).
16. John Leland, "Twilight of a Difficult Man: Larry Kramer and the Birth of AIDS Activism," *New York Times*, May 19, 2017 (accessed August 3, 2019).
17. Larry Kramer, "1,112 and Counting," *New York Native* 59 (March 14–27, 1983).
18. Mark Engler and Paul Engler, *This Is an Uprising: How Nonviolent Revolt Is Shaping the Twenty-First Century* (New York: Nation Books, 2016).
19. UNAIDS, "Fact Sheet: Global AIDS Update 2019" (accessed August 12, 2019).
20. U.S. Department of Health and Human Services, "Global Statistics," HIV.org (accessed October 23, 2019).
21. Sarah Schulman, *Let the Record Show: A Political History of ACT UP New York, 1987–1993* (New York: Picador/Farrar, Straus and Giroux, 2022).

22 Citation from *Rolling Stone*.
23. Deborah B. Gould, *Moving Politics: Emotion and ACT UP's Fight Against AIDS* (Chicago: University of Chicago Press, 2009).
24. Mark Engler and Paul Engler, *This Is an Uprising*.
25. Anthony Leiserowitz et al., *Climate Change in the American Mind: April 2019* (New Haven, CT: Yale Program on Climate Change Communication, 2019).
26. Anthony Leiserowitz et al., *Climate Change in the American Mind: September 2021* (New Haven, CT: Yale Program on Climate Change Communication, 2021 (accessed November 27, 2022).
27. Ibid., 17.
28. Good Grief Network, "Home: 10-Steps to Personal Resilience & Empowerment in a Chaotic Climate," November 2022 (accessed November 27, 2022).
29. Ibid., *Climate Café* (accessed November 27, 2022).
30. Mark U. Edwards, *Printing, Propaganda, and Martin Luther* (Berkeley: University of California Press, 1994).
31. Alexis Madrigal, "When the Revolution Was Televised," *The Atlantic*, April 1, 2018 (accessed August 18, 2019).
32. Sasha Torres, *Black, White, and in Color: Television and Black Civil Rights* (Princeton, New Jersey: Princeton University Press, 2003).
33. Monica Anderson, "3. The Hashtag #Blacklivesmatter Emerges: Social Activism on Twitter," *Pew Research Center*, August 15, 2016.

Step Five: Join the Movement and Disrupt Normalcy

1. Theda Skocpol, "Naming the Problem: What It Will Take to Counter Extremism and Engage Americans in the Fight Against Global Warming." Prepared for the symposium on The Politics of America's Fight Against Global Warming, Harvard University, 2013.
2. Rebecca Hersher, "Key Moments in the Dakota Access Pipeline Fight," *NPR*, February 22, 2017.
3. Floris White Bull, "I've Been Woken," *Awake: The Film* (accessed August 4, 2019).
4. Margaret Klein Salamon, "Venture Philanthropy for the Climate Emergency Movement: Funding Transformative Startups," *Climate Emergency Fund*, October 19, 2022.
5. Thomas Fortuna, "Plebeians Win Victory for the Rule of Law in Ancient Rome, 449 BCE. (See Also 494 Campaign)," *Global Nonviolent Action Database*, December 9, 2011.
6. Mark Engler and Paul Engler, *This Is an Uprising: How Nonviolent Revolt Is Shaping the Twenty-First Century* (Bold Type Books, 2016).

7. Ashley Ahn, "Protesters Interrupt Ted Cruz's Interview on 'The View,'" *NPR*, October 25, 2022.
8. Noam Scheiber, "How the Pandemic Has Added to Labor Unrest," *New York Times*, November 1, 2021.
9. Albert Einstein Institution, *198 Methods of Nonviolent Action*, December 2014.
10. Gene Sharp and Marina Finkelstein, *Power and Struggle: The Politics of Nonviolent Action: Part 1* (Porter Sargent, 1973).
11. "'I've Been to the Mountaintop' by Dr. Martin Luther King Jr." *AFSCME*, 2022.
12. Daniel Q. Gillion and Sarah A. Soule, "The Impact of Protest on Elections in the United States," *Social Science Quarterly* 99, 5 (2018): 1649–64.
13. Eryn Campbell, John Kotcher, Edward Maibach, Seth Rosenthal, and Anthony Leiserowitz, "Who Is Willing to Participate in Non-Violent Civil Disobedience for the Climate" (New Haven, CT: Yale Program on Climate Change Communication, 2022).
14. Mark Engler and Paul Engler, *This Is an Uprising*.
15. Casey Quackenbush, "After Two Years, Extinction Rebellion Returns," *Intelligencer*, April 15, 2022.
16. Kim Klein, *Fundraising for Social Change* (San Francisco: Jossey-Bass, 2011).
17. Paul Engler, "Protest Movements Need the Funding They Deserve," *Stanford Social Innovation Review*, July 3, 2018 (accessed August 12, 2019).

Conclusion: All-In for All Life

1. Abraham Maslow, "A Theory of Human Motivation," *Psychological Review* 50 (4), 370–96.
2. Abraham Maslow and Rober Frager, *Motivation and Personality* (New Delhi: Pearson Education, 2007).
3. Mark E. Koltko-Rivera, "Rediscovering the Later Version of Maslow's Hierarchy of Needs: Self-Transcendence and Opportunities for Theory, Research, and Unification," *Review of General Psychology* 10, 4 (2006): 302–317.
4. Kim Polman and Stephen Vasconcellos-Sharpe, *Imaginal Cells; Visions of Transformation* (London: Reboot the Future, 2017).

INDEX

1,112 and Counting (Kramer), 85–86

A

acceptance of feelings, 42–43
Activate Hope: How to Face the Mess We Are in Without Going Crazy (Macy), 56
AIDS Coalition to Unleash Power (ACT UP), 83–87, 89
alienation, 67
all-in for all life, 5, 64, 71–72, 107, 118
altruism, 71
amnesia, 44
apathy, 44
Art of Loving, The, (Fromm), 3
avoiding the truth, 16
awakening, collective, 6, 12, 75, 93, 102, 115

B

back-to-basics approach, 28
Baines, Jessie-Ann, 90
Bertini, Chloe, 117
Black Lives Matter, 93
Blitz, 78
Blumenauer, Earl, 82
Brach, Tara, 48
bridge-blocking action, 98–100
Brown, Brené, 67
Buddhism, 59, 72
bunkers, building, 21
Butler, Octavia, 55
butterfly, 118–19

C

capitalism, 7
catastrophic scenarios, 24
caterpillar, 118–19
change, transformative, 15, 23, 68, 96, 102
Chong, Mun, 106
Chouniard, Ivan, 77
citizens in a democracy, 37
civil
  disobedience, 95, 100–101, 104
  resistance, 98
  rights, 38
Civil Rights Movement, 92
civilizational stabilizing mechanism, 28
Climate Café, 91
climate "doomers", 58–59
Climate Emergency, city-based, 10
Climate Emergency Fund (CEF), 11–12, 82, 96–97, 103, 111
climate emergency movement, 39–40, 95
Climate Emergency Unit, 32, 77
climate emotions, 60, 67
Climate Emotions Conversations, 10, 90–91
climate legislation, 98–99
Climate Mobilization, The (TCM), 9, 10, 82, 99
Climate Movement 2.0, 96
Climate Movement 3.0, 97

climate philanthropy, 97–99
Climate Psychologist, The, (Klein Salamon), 9
collective denial. *See* denial, collective
collective
  response, 32, 83
  solution, 89
communications technologies, cutting edge, 92
compartmentalization, 21
compassion, 28
confronting emotions, 31
connection, 28, 59–61, 68, 71, 73, 100
consumer-capitalist society, psychological impacts of, 4
contributing to the movement, 13
COVID-19, 35–36
Cruz, Ted, 101
curiosity, 46–47

D

Dakota Access Pipeline, 96
death as sweet relief, 53
Declaration of Climate Emergency, 82, 98–99
denial
  campaign, 17–18, 21, 39
  collective, 36, 39, 52–53, 67, 6, 93
  of emotions, 41–42
  peace of, 16, 24, 29, 61
Dernière Renovation, 12
destruction
  appropriateness of, 5
  internalization of, 5
destructive forces, 72
discussing global warming, 89–90
disruptive action, 12, 40, 96–98, 102–104
dissociation, 22, 44, 46
donations, 109–111, 113
drawdown, 8, 25, 28, 40, 80
DREAMers, 89
Dunlop, Ian, 22

E

Eaarth (McKibben), 30
eco-apartheid, 20
ecological footprint, reduction of, 81
economy, regenerative, 8
emergency
  mobilization, 16, 32–35, 7–80, 82, 85–87
  mode, 75–80, 83–84, 89, 110
emotional muscle, 43, 47
end of all life, possibility of, 5
Engler, Mark and Paul, 86, 100
equality, gender and racial, 34
Europe heatwave (2022), 25
existential risk, 17, 22, 118
Extinction Rebellion (XR), 10, 19, 76, 98–99, 105–106
Exxon Mobil, 3, 17

F

fair distribution, 35
false self, 4
fear
  emotion of, 54–55, 89
  as motivator, 22–23, 50–51, 54
"fear of fear", 23
feel your feelings, 41–52
financial sacrifices, 78, 109–11
food insecurity, 27
food provision, 28
flooding Pakistan (2022), 24, 27
fossil fuel companies,
  denial campaign, 17–18, 21
  responsibility of, 3

Francis, Pope, 8, 46
Fromm, Erich, 3–5, 13
fugue states, 44
fundraising, 99–100, 112-114
Fundraising for Social Change (Klein), 112

G

Gay Men's Health Crisis (GMHC), 84
Getty, Aileen, 100
Gilbert, Daniel, 83
Gilding, Paul, 26
Giving Green, 98
global economy, 81
Global Footprint Network, 26
Global North, 27
Global South, food insecurity of, 27, 56
Good Grief Network, 91
Gore, Al, 23
governmental failure, 3
"gradualist" climate movement, 22–23
Great Disruption, The, (Gilding), 26
Green New Deal, 81
grief, emotion of, 54–62
group selection, 70–71

H

Haq, Zain, 63–64
Havel, Václav, 38
Heartland Institute, 18
heatwave Europe (2022), 25
Heglar, Mary Annaïse, 57
hierarchy of needs, 116–17
HIV 83–87
Holdfast Collective, 77
hope as motivator, 60–61
housing provision, 28
human
  needs, 7, 82
  potential, immensity of, 6, 51
humanity's basic needs, 28
humans as stewards of nature, 7
Hurricane
  Ian, 25
  Sandy, 9

I

ignorance
  pluralistic, 31, 38
  willful, 21, 29, 44
Imaginal Cells: Visions of Transformation (Polman), 118
Inconvenient Truth, An, (Gore), 23
Indigenous water protectors, 19, 96–97
individual selection, 70–71
individualism, extreme, 60
Inflation Reduction Act (IRA), 19–20, 98
instinct, psychological, 2
intellectualization, 21
interconnectedness, 60, 71–72
Intergovernmental Panel on Climate Change (IPCC), 17, 22, 25
interrelated factors, 27

J

Jim Crow laws, 37, 92
Jones, Mother, 59
journaling, 48–49
Junger, Sebastian, 78
Justice Democrats, 81
Just Stop Oil, 12, 76, 95

K

Kaiser, David, 78
Kelly, Kerry, 43

Kennedy, Rory, 100
King Jr., Martin Luther, 71, 102, 105
Klein, Kim, 112
Kramer, Larry, 46, 84–86

L

Lancet Planetary Health, The, xiii, 2
Laudato Si' (Pope Francis), 8
learning from past experiences, 63–69
Lent, Jeremy, 70
Letter from a Birmingham Jail (King Jr.), 71
listening, 88
Living Planet Report 2016 (World Wildlife Fund), 57
Living Planet Report 2018 (World Wildlife Fund), 26
living the climate truth, 11, 16, 44–45, 57, 60, 82
love as motivator, 60–61
Love of Death and Love of Live (Fromm), 4
Luther, Martin, 92

M

Macy, Joanna, 56, 59–60, 91
manipulation, political, 18
Mann, Michael, 22
Maslow, Abraham, 116–17
mass movement, 76
McKibben, Bill, 30, 88
meditation, 48
Merchants of Doubt (Oreskes), 17
meritocracy, 37
#MeToo, 37, 92
mindfulness, 48
minimization of climate emergency, 22
mobilization of resources, 16, 25, 81–82
moral
  courage, 28
  obligation, 109, 114
movement mentality, 11, 64, 73, 77, 88
multiple personality disorder, 44

N

Nakate, Vanessa, 60
national purpose, 34
Neff, Kristin, 43
negentropy, 70, 73
Neilson, Trevor, 100
Nongovernmental International Panel on Climate Change (NIPCC), 18
nonviolent action, 101–2, 104–5
Nonviolent Direct Action (NVDA) Database (Swarthmore College), 101
Novick, Jack, 43

O

Ocasio-Cortez, Alexandria, 81–82, 96–97
Occupy Wall Street, 101
Oreskes, Naomi, 17
out-of-body experiences, 44
Overton window, 23

P

pain
  of the climate crisis, 30
  discomfort of, 44
Pakistan flooding (2022), 24, 27
Parable of the Sower (Butler), 55
Paradise Built in Hell, A, (Solnit), 79
Paris Agreement (2015), 25
participation in a destructive system, 2

passive
acceptance, 5
ignorance, 12
Patagonia, 77
Pearl Harbor, 83
people as commodities, 4–5
personal
responsibility, 9, 115, 119
suffering, 8
pipeline fighters, 19, 96
Planet3.0, 24
Plebeian activists, 100–101
pluralistic ignorance, 31, 38
policy recommendations, 81
Polman, Kim, 118
power of truth, 37–39
powerlessness, myth of, 37
processing feelings, 43
productive cooperation, 88
projection, 21
Protestant Reformation, 92
psychoanalysis, 45–46
psychotherapy, 46, 102
"pull-together effect", 78

R

RAIN, 48
rationing program, 35
reality, reckoning with, 15, 55–56
regression, 21
responsibility of fossil fuel companies, 3
Rise Up Movement, 60
risk tolerance, 105–106
Rushkoff, Douglas, 20

S

"sacrifice zones", 27
Sanders, Bernie, 82
"scholarly reticence", 22
School Strikers, 19, 76
Schrödinger, Erwin, 70
Schulman, Sarah, xiii, 87
Scientist Rebellion, 76
second law of thermodynamics, 70
self-actualization, 117
self-care, 49
self-compassion, 43, 47, 49, 55
self-criticism, 88
self-interested individuals, 37, 87
self-judgment, 42, 47, 50
self-reflective change process, 43
self-transcendence, 117
shame, 67–68
Silence = Death, 84–85
Silk, Ezra, 9, 80, 82
sixth mass extinction, 50, 81
skills for volunteering, 108–9
Social Conquest of Earth, The, (Wilson)
social
contract, 3
media, 18
movement, 9, 38, 45–46, 84, 112–113
socialism, resurgence of, 7
Solnit, Rebecca, 79
spiritual
birth right, 115
sense, 69
Spratt, David, 22
Standing Rock, 96–97
State of Food Security and Nutrition in the World (UN), 27
suffering, 56
Summary for Policymakers (IPCC), 22
Sunrise Movement, 19
superpowers, 64, 105
Survival of the Richest: Escape Fantasies of Tech Billionaires (Rushkoff), 20

Swarthmore College, 101

T

talk about climate, 89–92
tax increases 6, 35
TCM. *See* Climate Mobilization, The
Thatcher, Margaret, 37
therapy (psycho), 45–46
thermodynamics, second law of, 70
Thich Nhat Hanh, 59
This Is an Uprising (Engler), 86, 100
thought leadership, 10, 98
time commitment, 106–8
tipping points, climate, 20
"tobacco strategy", 17
Tobis, Michael, 24
Tour de France, 12
transformative change, 15, 23, 68, 96, 102
traumatic impact of psychotic episodes on romantic partners, 67
true security, 110
truth, power of, 37

U

understating risks of climate emergency, 22
Uninhabitable World, The, (Wallace-Wells), 16–17

V

Velvet Revolution, 38
Victory gardens, 81
Victory Plan (Silk), 80–82
View, The, 101

W

Wallace-Wells, David, 15–17, 22
wealth inequality, 117
What Is Life (Schrödinger), 70
White Bull, Floris, 97
White House, 12
Why Scientists Disagree About Global Warming (Heartland Institute), 18
wildlife corridor, global, 81
willful ignorance, 21, 29, 44
Wilson, E.O., 70–72
window of political discourse, 23–24
wishful thinking, 21
Word of the Year, 10
Work That Reconnects Network, 91
World Wildlife Fund, 26, 57
WWII-scale mobilization, 10, 32–35, 78, 81–82, 110

Z

Zen Buddhism, 59
zero CO_2 emissions, 25, 28, 39–40, 80
"zoning out", 44

ABOUT THE AUTHORS

PHOTO: AIMEE ALMSTEAD

MARGARET KLEIN SALAMON, PHD, is a clinical psychologist turned climate activist. She is obsessed with waking people up to the climate emergency, enlisting them in the cause of protecting humanity and all life, and galvanizing a more urgent and effective climate movement. She has pursued this mission as a thought-leader, a fundraiser, an organization-builder, and currently as the Executive Director of Climate Emergency Fund. She lives in Brooklyn with her husband and two dogs, Hero and Cassandra. She loves her life and doesn't want to die in a climate apocalypse.

MOLLY GAGE, PHD, is a book developer and specializes in transformative nonfiction.

ABOUT NEW SOCIETY PUBLISHERS

DON'T EAT THIS BOOK (but you could)

New Society Publishers is an activist, solutions-oriented publisher focused on publishing books to build a more just and sustainable future. Our books offer tips, tools, and insights from leading experts in a wide range of areas.

We're proud to hold to the highest environmental and social standards of any publisher in North America. When you buy New Society books, you are part of the solution!

At New Society Publishers, we care deeply about *what* we publish—but also about *how* we do business.

- This book is printed on **100% post-consumer recycled paper,** processed chlorine-free, with low-VOC vegetable-based inks (since 2002).
- Our corporate structure is an innovative employee shareholder agreement, so we're one-third employee-owned (since 2015)
- We've created a Statement of Ethics (2021). The intent of this Statement is to act as a framework to guide our actions and facilitate feedback for continuous improvement of our work
- We're carbon-neutral (since 2006)
- We're certified as a B Corporation (since 2016)
- We're Signatories to the UN's Sustainable Development Goals (SDG) Publishers Compact (2020–2030, the Decade of Action)

To download our full catalog, sign up for our quarterly newsletter, and to learn more about New Society Publishers, please visit newsociety.com

ENVIRONMENTAL BENEFITS STATEMENT

New Society Publishers saved the following resources by printing the pages of this book on chlorine free paper made with 100% post-consumer waste.

TREES	WATER	ENERGY	SOLID WASTE	GREENHOUSE GASES
16	1,300	7	55	6,990
FULLY GROWN	GALLONS	MILLION BTUs	POUNDS	POUNDS

Environmental impact estimates were made using the Environmental Paper Network Paper Calculator 4.0. For more information visit www.papercalculator.org

Certified B Corporation

new society PUBLISHERS www.newsociety.com

FSC www.fsc.org MIX Paper from responsible sources FSC® C016245

SDG PUBLISHERS COMPACT